UP FROM NOTHING

UP FROM NOTHING

THE MICHIGAN STATE UNIVERSITY CYCLOTRON LABORATORY

SAM M. AUSTIN

foreword by LOU ANNA K. SIMON
President, Michigan State University

MICHIGAN STATE
U N I V E R S I T Y

Printed and bound in the United States of America.

♾ The paper used in this publication meets the minimum requirements
of ANSI/NISO Z39.48-1992 (R 1997) (Permanence of Paper).

ISBN: 978-0-99672-520-0 (cloth)
ISBN: 978-0-99672-521-7 (paper)
ISBN: 978-0-99672-522-4 (ebook: PDF)

19 18 17 16 15   1 2 3 4 5

Designed and typeset by Charlie Sharp, Sharp Des!gns, Lansing, MI

CONTENTS

FOREWORD

Lou Anna K. Simon

President, Michigan State University

Michigan State University is fortunate, indeed, that University Distinguished Professor Emeritus Sam M. Austin has dedicated his time and exceptional narrative skill to provide us with an authoritative and engaging account of the cyclotron era at MSU.

Like many young physicists who earned their doctorates in the two decades following World War II, he was drawn to the emerging field of nuclear physics as representing the frontier of knowledge. At that time, the entire nation was both excited and apprehensive about the potentials of nuclear science.

The power of the atom's nucleus, uncovered by scientists in the 1930s, had delivered victory to the Allies, bringing World War II to a precipitous end. Although this explosive power proved in time to be a mixed blessing, the magnitude of energy it displayed also held promise to serve society in peacetime, offering a new source of energy to generate electricity for communities and, with more research and development, to yield solutions to other of society's needs.

President John Hannah, a leader who was always looking around the curve, picked up on an idea advanced by the scientist-administrators around him that there would be a rich future for nuclear physics research and that nuclear physics could be a good fit for Michigan State. At the time, although relatively unrecognized, MSU had a strong cadre of scientists, particularly in the life sciences—the study of living organisms at the cellular and molecular level. Hannah and his colleague administrators postulated that low-energy nuclear physics—the study of the structure and stability of nuclei in the world around us and the cosmos, in other words the study of the nature of matter—could align perfectly, and in time double MSU's research strength and stature in the natural sciences among the nation's major public universities.

Yes, it was a bold risk, but Hannah and his leadership team took it. And thus MSU started on what seemed to many to be an improbable journey to the frontier of nuclear science. This journey has carried us "up from nothing" to the creation of the first K50 cyclotron in the 1960s and, today, to the creation of the Facility for Rare Isotope Beams. Few would have guessed, fifty years ago, that a university lacking infrastructure or prowess in nuclear physics research would eventually develop a progression of "discovery machines" that would make it a destination for nuclear scientists from around the world.

The account Professor Austin relates in this book is a tale emblematic of the history of Michigan State University. One hundred sixty years ago, MSU pioneered a bold new experiment in learning. What began as a college dedicated to serving Michigan communities, chartered to democratize classical education and couple it with the application of practical knowledge, also grew in time "up from nothing" to become a leading international research university.

The rise of both MSU and nuclear science at MSU exemplify cultural attributes both share: courage, hard work, confidence, and determination to be the best. Today, as the Facility for Rare Isotope Beams rises from its foundations to lead rare isotope research tomorrow, the cyclotrons continue their forefront experiments to lead the science today.

The coupling of a national user facility for nuclear science and a leading university grounded in research is a logical union. The global evolution of modern scientific research has placed universities in crucial positions as engines of innovation, driving long-term investigations into solutions for society's biggest problems. The result of this union between university and user facility is synergistic, bringing together the best minds and the right tools, so the whole can become greater than the sum of its parts. The relationship is a unique and fruitful one.

The knowledge generated here can be developed and built into businesses in emerging fields to satisfy both pressing and unforeseen societal needs. As the world leader in rare isotope science, our nuclear scientists and the users of our facilities have continued to push back the frontiers of knowledge and produce important breakthroughs as well as applications of isotopes in medicine, materials research, national security, and even agriculture. The discoveries are both theoretical and practical.

The bold confidence of the cyclotron era developed a culture intent on delivering results, where a mindset of constant, deliberate reinvention responded to the aspirations of nuclear scientists by designing and building marvels of experimental equipment. Resilient leadership and inventive, collaborative minds accelerated this culture into the twenty-first century. The Department of Energy's entrusting MSU to deliver the FRIB Laboratory is, at once, a symbol of how far we have come and a continued promise of future breakthroughs.

To stay at the leading edge of education and discovery, we must align ourselves with the tireless pursuit of innovation, taking the rise of nuclear science at MSU as our example. In the pages ahead, Professor Austin guides us through the challenges and the triumphs of difficult tasks ultimately well done, from which we can draw lessons and take heart. He shows us the power of building strong programs through supporting extraordinary people and how generating aggregations of creative intellects can advance our understanding of complex problems.

This story of our bold predecessors provides us with an example of how we can approach other great problems in our society. How can we find solutions to our long-term energy needs? How can we provide water resources to an agriculture system that can feed the populations of the world effectively? How can we live together in a world of sustainable prosperity? Through continuing our faith in curiosity, our culture of commitment, and our dedication to the courageous pursuit of big ideas, we can embark on more improbable journeys and answer these questions together.

The innovators who came before the rise of FRIB over the past half-century took great chances and accepted great risks. When they failed, they approached each problem again with renewed vigor and confidence and never gave up until they achieved excellence.

Faculty and researchers come to MSU with big ideas that they want to bring to life, to achieve radical and tangible applications. Students come to MSU with big dreams they want to make bigger. I know that *Up from Nothing*, as the story of the emergence of world-class nuclear science research at MSU, will provide some guidance toward achieving what seem like difficult, even improbable goals and illuminate the roles that persistent, forward-looking people and persistent, forward-looking institutions play in advancing knowledge and transforming lives.

ABOUT THE AUTHOR

I grew up on a general purpose farm not far from Madison, Wisconsin, and began my education in the Goebel School, a one-room school with one teacher, eight grades, and twenty-five students, followed by Columbus High School in a class of about seventy.

Attending college was uncommon in my rural community, and I intended to follow in my farmer father's footsteps. I was very active in Future Farmers of America (FFA) and took few college preparatory courses: no biology at all and mathematics only through geometry. But after completing high school in June 1951, I somehow realized that family farming was a dying profession.

That summer, at the last minute, I applied for entrance to the University of Wisconsin–Madison (UW) as a mathematics major and was admitted with a scholarship UW gave to high school class valedictorians.

As an undergraduate at UW, I worked with physics professor Ray Herb and received thorough training in experimental physics. We tried to build a high precision ion source using field-emitted ions, but it failed because Herb hadn't realized, or had forgotten to tell me, that palladium lacked sufficient tensile strength to withstand the strong electrical forces; famous physicists can be wrong.

I stayed at UW for graduate school, where I led two experiments using neutrons. I then went to the University of Oxford in the UK on a National Science Foundation Postdoctoral Fellowship and next accepted a faculty position at Stanford University.

The Stanford experience greatly affected my viewpoint of how laboratories should be run. The Stanford physics department was dominated by the views of Felix Block, a Nobel Prize winner of exceptional talent who believed the nature of a physics department should be determined by very few senior Nobel Prize winners who would make all important decisions. These decisions were sometimes flawed.

For example, when the construction of SLAC, the Stanford Linear Accelerator Center, was proposed, the physics department did not happily support it. SLAC has since contributed greatly to the advancement of particle physics. Nor were two of my assistant professor colleagues, who later won Nobel Prizes, viewed as good enough for Stanford. Others who happened to work closely with the Nobel winners achieved tenure, but later they contributed much less to science.

I did well enough to receive a Sloan Research Fellowship and professorship offers from institutions that had strong nuclear science programs. My experience at Stanford led me to consider issues other than reputation, and I accepted the offer from little-known Michigan State University (MSU). I thought that its new Cyclotron Laboratory was run by people driven by science, and I felt it offered the best opportunity for young and ambitious faculty. This has proven to be the right decision for me and, I believe, for many others of similar mind who came to MSU. When I eventually held leadership positions in the laboratory, I tried to maintain the mindset that influenced me to come to MSU.

As you can see, I am not trained as a historian. However, I have the advantages of being alive and active in nuclear physics, and on the scene at MSU for all but the first few years of the Cyclotron Laboratory's existence. Consequently, I have firsthand knowledge and insight into decisions and dynamics that might be difficult for a later writer to obtain. I hope that my account can provide useful background for any later and more detailed accounts.

Time has passed, however, and not all of the participants active in the development of nuclear science at MSU, from 1955 to 1985, were available to provide their viewpoints. General information on this period is provided in two histories: one written for the 1955 centennial of the founding of MSU and the other written for MSU's sesquicentennial in 2005. For additional detailed information about this period, I have relied primarily on a synthesis of the somewhat fragmentary records available in the MSU Archives and in the records of the MSU Department of Physics and Astronomy; the

Cyclotron Laboratory; the National Superconducting Cyclotron Laboratory (NSCL); first laboratory director Henry Blosser's papers; and my own records. On occasion I have relied on my memory of events. These various documents and other sources are cited in the text and enumerated in the bibliography at the end of the document.

I believe that these earlier years in the Cyclotron Laboratory's history established the culture, attitudes, and practices that have led to its ongoing success. I have, therefore, emphasized these years in this account, dealing with more recent events in somewhat less detail. I have, however, tried to convey the essence of the activities that have determined the evolution of the laboratory. Fortunately, for those desiring a more detailed description of these years, information on the laboratory's activities is widely available in annual reports and proposals.

▶ To meet the needs of audiences with more or less interest in technical or organizational details, I have made extensive use of insets (denoted by a green heading) and appendices with the goal of producing a more readable main text. A few of the endnotes are relatively extensive or informative; their numbers are in **boldface**.

In later years, as the Cyclotron Laboratory gained national stature and was awarded a growing portion of U.S. funding for nuclear science, the laboratory's activities were more exposed to political considerations. The political atmosphere became particularly intense in 1999 and the years following when the laboratory set its sights on attaining for MSU the Facility for Rare Isotope Beams (FRIB) that will be the major U.S. nuclear physics facility developed for the twenty-first century. I've outlined the course of events that led to the U.S. Department of Energy's (DOE) 2008 choice of MSU to build FRIB and the initial project activities since then. But a detailed description of the ongoing project development that is expected to deliver FRIB in 2022 must await future exposition.

Finally, over the years, hundreds of individuals have dedicated their talents and skills to creating the ongoing success of the Cyclotron Laboratory at MSU. I appreciate all their contributions and regret that I could not give them all the individual credit they deserve for their roles in advancing the laboratory and nuclear science. Partly this owes to my need to achieve a suitably short document, partly to the nature of my experience and lack of exposure to the entire array of activities in the laboratory, partly to vagaries of memory and availability of records and partly to my decision to emphasize the role that accelerator development has played in the evolution of the laboratory.

Finally, I note that much of the detailed information provided here was collected by early 2014, and numbers quoted usually refer to late in 2013.

GLOSSARY

■ **U.S. Funding Agencies for Nuclear Science**

AEC. Atomic Energy Commission.

DOE. Department of Energy.

DOE-SC. Department of Energy Office of Science.

ERDA. Energy Research and Development Agency.

NSAC. Nuclear Science Advisory Committee, provides advice to DOE and NSF regarding nuclear science issues, including funding.

NSF. National Science Foundation.

■ **U.S. Department of Energy Nuclear Science Laboratories/Facilities**

ANL. Argonne National Laboratory, a National Laboratory located in Argonne, Illinois, near Chicago. It has a strong nuclear physics activity using a superconducting linear accelerator.

ANL was a major competitor to MSU in the effort to secure the Facility for Rare Isotope Beams (FRIB). Est. 1946.

BNL. Brookhaven National Laboratory, a National Laboratory located on Long Island, New York. It has a strong nuclear physics activity using the Relativistic Heavy Ion Collider (RHIC). Est. 1948.

FRIB. Facility for Rare Isotope Beams at Michigan State University.

JLAB. Thomas Jefferson National Accelerator Facility, an electron accelerator facility located in Newport News, Virginia. It specializes in studies of the structure of very light nuclei, especially the proton. Est. 1984.

LBNL. Lawrence Berkeley National Laboratory, a National Laboratory located in Berkeley, California. The cyclotron was invented by E. O. Lawrence in Berkeley, and the laboratory now bears his name. Est. 1954.

LLNL. Lawrence Livermore National Laboratory, Livermore, California. Est. 1952.

ONRL. Oak Ridge National Laboratory, Oak Ridge, Tennessee. Est. 1943.

■ General Terms

A. Mass number = N + Z, the number of nucleons (neutrons + protons) in a nucleus.

alpha particle (α). The nucleus of the helium atom, it contains two neutrons and two protons. Alpha particles emitted by decaying heavy nuclei were used in early nuclear physics experiments.

apparatus. A device or devices used to make measurements.

atom. A microscopic entity consisting of a compact nucleus, composed of neutrons and protons, surrounded by an electron cloud 100,000 times as large.

atomic mass unit (amu). 1/12 the mass of the ^{12}C nucleus, which has Z = N = 6.

atomic number. The number of protons in the nucleus of an element.

atomic weight. The mass of the average nucleus for an element compared to the atomic mass unit (amu). Usually the atomic weight is, within about 0.1 percent, equal to A = N + Z.

detection equipment. A device or devices used to detect products of nuclear reactions.

deuteron (d). The nucleus of the heavy hydrogen atom. Consists of a proton and a neutron.

electron. A negatively charged particle with mass about 1/2,000 that of the proton. It forms a cloud of negative charge around the nucleus of an atom.

element. An entity with a given atomic number, or number of protons in its nucleus.

experimental physics. Study of nature by making measurements (i.e., by performing experiments) and trying to understand their implications.

heavy ion. Not precisely defined, but usually assumed to be an ion heavier than an alpha particle.

ion. An atom with some number of electrons removed, therefore positively charged.

isotopes. An isotope is one of the possible types of an element. All isotopes of a given element have a nucleus with the same number of protons but different numbers of neutrons.

N. Number of neutrons in a nucleus.

neutron (n). An uncharged particle that is one of two main components of nuclei. It is radioactive and has a mass slightly greater than that of a proton.

nuclear mass. The mass of a nucleus compared to the atomic mass unit (amu), which is 1/12 the mass of the ^{12}C nucleus.

nuclear science. Experimental or theoretical study of the properties of nuclei and nuclear matter.

nucleon. A neutron or a proton.

nucleus. A compact system consisting of neutrons and protons located at the center of an atom. It contains almost all the mass of the atom (>99.9 percent).

nuclide. A nucleus with a particular number of neutrons, N, and a particular number of protons, Z.

particle physics. Experimental or theoretical study of the properties of particles. Some are known to have substructure (e.g., protons and neutrons, which contain quarks and gluons) and others do not (e.g., neutrinos, electrons).

proton (p). One of the two main components (the neutron is the other) of nuclei. It has electric charge and is the nucleus of the hydrogen atom.

rare isotope. An uncommon isotope with a particular number of neutrons, N, and a particular number of protons, Z. In past usage, this would have been called a rare nuclide. Rare isotopes are usually radioactive.

theoretical physics. Study of nature by using calculations to see which models agree with experimental results, and then predicting related phenomena.

Z. Number of protons in a nucleus.

■ Specialized Terms

A1200. First NSCL fragment separator. The "A" stands for analysis and the "1200" for bending power.

A1900. Current NSCL fragment separator. The "A" stands for analysis and the "1900" for bending power.

ALARA. As low as reasonably achievable, referring to radiation exposure.

ARTEMIS. Advanced Room Temperature ECR Ion Source at NSCL.

ATLAS. Argonne Tandem Linac Accelerator System at ANL.

BCS. Beta Counting System, series of silicon detectors to study beta decay.

BECOLA. Facility for laser spectroscopy and beta-NMR.

Beta-NMR. Beta-Nuclear Magnetic Resonance Apparatus for measuring magnetic moments.

Bonner Prize. Tom W. Bonner Prize in Nuclear Physics, an award of the American Physical Society.

BigRIPS. Big RIKEN Projectile Fragment Separator, at RIBF.

BUU. Bolzmann-Uhling-Uhlenbeck approximation, for collisions of heavy ions.

CCF. Coupled Cyclotron Facility at NSCL.

CEBAF. Continuous Electron Beam Accelerator Facility at JLAB.

CENPA. Center for Experimental Nuclear Physics and Astrophysics at University of Washington.

CERN. European Organization for Nuclear Research.

Chandra. Satellite X-ray Observatory.

DRAGON. Detector of Recoils and Gammas from Nuclear Reactions at TRIUMF.

DWBA. Distorted Wave Born Approximation.

EBIS. Electron Beam Ion Source.

EBIT. Electron Beam Ion Trap.

EC. Electron Capture.

ECR. Electron Cyclotron Resonance.

ECRIS. Electron Cyclotron Resonance Ion Source.

EOS. Equation of State.

FAIR. Facility for Antiproton and Ion Research (GSI).

Fermilab. Fermi National Accelerator Laboratory, Batavia, Illinois.

FNAL. Fermi National Accelerator Laboratory, a particle physics laboratory.

FRS. Current fragment separator at GSI.

GANIL. Grand Accelerateur National d'Ions Lourds, Caen, France, a nuclear physics laboratory.

GDR. Giant Dipole Resonance.

GMR. Giant Monopole Resonance.

GRETA. Gamma-Ray Energy Tracking Array.

GRETINA. First stage of GRETA, built mainly at LBNL.

GSI. Gesellschaft fur Schwerionenforschung, Darmstadt, Germany, a nuclear physics laboratory.

GT. Gamow-Teller, beta decay or capture with zero orbital angular momentum.

HIRA. High Resolution Array Detector, Array of twenty particle telescopes, Si + CsI at NSCL.

HRIBF. Holifield Radioactive Ion Beam Facility at ORNL.

INT. Institute for Nuclear Theory, at University of Washington.

IPN Orsay. L'institut de physique nucléaire d'Orsay.

ISAC. Isotope Separator and Accelerator at TRIUMF.

ISOL. Isotope Separation On-Line.

ISOLDE. Isotope Separator On-Line at CERN, Geneva, Switzerland.

JINA. Joint Institute for Nuclear Astrophysics.

JINA-CEE. Joint Institute for Nuclear Astrophysics, Center for the Evolution of the Elements at MSU.

K. A measure of the bending power of a magnet system, in MeV for protons.

K50, K100, K500, K1200. Cyclotron accelerators at NSCL.

KVI. Kernfysisch Versneller Instituut, Groningen, The Netherlands, a nuclear physics laboratory.

LEBIT. Low Energy Beam and Ion Trap facility for mass measurements at NSCL.

LENDA. Low Energy Neutron Detector Array at NSCL.

Liquid hydrogen target. Dense source of hydrogen for nuclear reaction studies at NSCL.

Linac. Linear accelerator.

LISA. Large Institutional Scintillator Array at NSCL.

Miniball. A compact large solid angle detector at NSCL.

MoNA. Modular Neutron Array at NSCL.

MoNA-LISA. Combination of MoNA and LISA.

MSU. Michigan State University, East Lansing, Michigan.

NAS. National Academy of Sciences.

NASA. National Aeronautics and Space Administration.

NERO. Detector for low energy neutrons at NSCL.

Neutron Walls. Two large-area, position-sensitive neutron detectors with neutron, gamma-ray discrimination at NSCL.

NMR. Nuclear Magnetic Resonance.

NNSA. National Nuclear Security Administration.

NSAC. Nuclear Science Advisory Committee.

NSCL. National Superconducting Cyclotron Laboratory, East Lansing, Michigan, a nuclear science laboratory at MSU.

NSSC. Nuclear Science and Security Consortium.

NuPECC. Nuclear Physics European Collaboration Committee.

PAC. Program Advisory Committee.

PAN. Physics of the Atomic Nucleus, JINA outreach program at NSCL.

Particle telescope. A device having two or more detectors, and combining their signals to identify particle types.

R&D. Research and Development.

RCNP. Research Center for Nuclear Physics at Osaka University.

REU. Research Experience for Undergraduates.

REX-ISOLDE. Radioactive beam Experiment (reaccelerator) at ISOLDE/CERN.

RF. Radio frequency.

RFFS. RF Fragment Separator to separate isotopes by velocity and purify p-rich beams at NSCL.

RFQ. Radio Frequency Quadrupole Accelerator.

RHIC. Relativistic Heavy Ion Collider at BNL.

RIA. Rare Isotope Accelerator.

RIB. Radioactive Ion Beam.

RIBF. Radioactive Ion Beam Factory at RIKEN.

RIKEN. The Institute of Physical and Chemical Research, Wako, Japan, a general purpose laboratory.

RIPS. Riken Projectile Fragment Separator at RIKEN.

S800. S800 Spectrograph at the NSCL. The "800" is the maximum bending power.

SC. Superconducting.

S-DALINAC. Superconducting Darmstadt Linear Accelerator at Darmstadt.

SEETF. Single Event Effects Test Facility, an in-air irradiation station for effects of radiation on electronics and other systems.

SeGA. Segmented Germanium Array at NSCL.

SHIPTRAP. Separator of Heavy Ion Reaction Products TRAP at GSI.

SLAC. Stanford Linear Accelerator Center, a particle physics and astrophysics laboratory.

SNS. Spallation Neutron Source at ORNL.

SRF. Superconducting Radio Frequency.

STAR. Solenoidal Tracker at RHIC.

SuN. Summing NaI detector for gamma ray total absorption measurements at NSCL.

SuperFRS. Super Fragment Separator at GSI built as part of FAIR.

SuSI. Superconducting ECR Source for Ions at NSCL.

Sweeper. Magnet used with MoNA-LISA, bends protons away from neutron detectors at NSCL.

TOF. Time of Flight.

TPC. Time Projection Chamber.

TRIPLEX. Nuclear lifetime measurements by recoil distance method (RDM) at NSCL.

TRIUMF. TRI-University Meson Facility, Vancouver, Canada, a nuclear and particle physics laboratory.

VENUS. Versatile ECR ion source for Nuclear Science at LBNL.

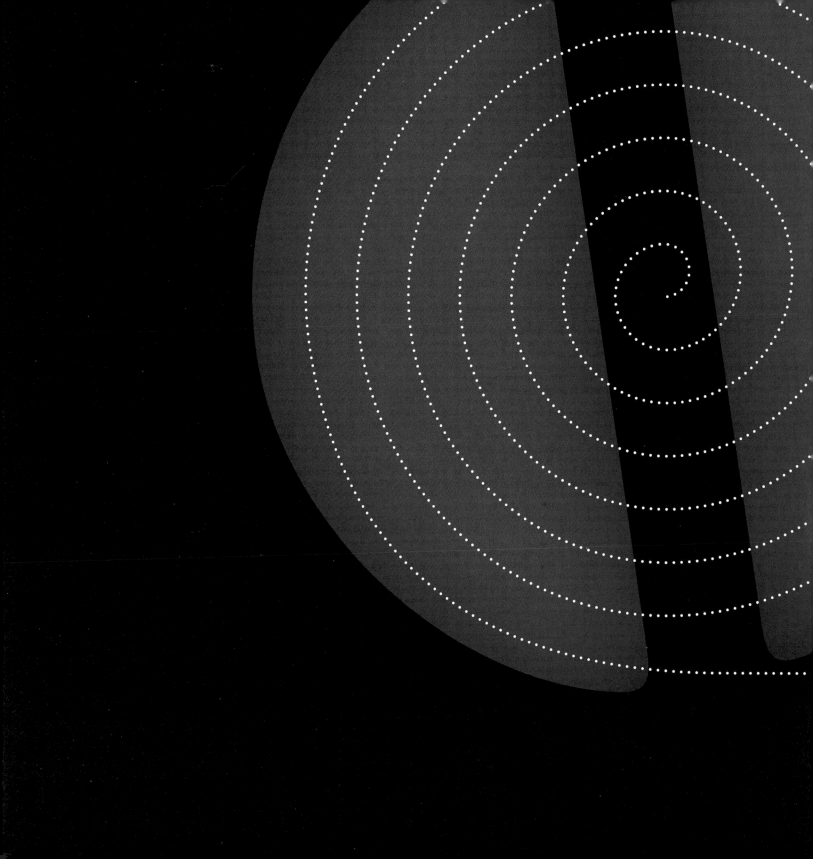

UP FROM NOTHING

FRIB: The Improbable Adventure

On December 11, 2008, Michigan State University's nuclear science team was chosen by the U.S. Department of Energy (DOE) to design and establish the Facility for Rare Isotope Beams (FRIB), the newest DOE national user facility, or as DOE calls it: a "discovery machine."

When completed, FRIB will be the world's most powerful accelerator for the production of rare isotopes. Many of these isotopes have a very fleeting existence, yet they will allow scientists to study the properties of nuclei in exquisite detail and to determine the role they play in the cosmic origin of the elements. Applications of these isotopes promise to benefit mankind and advance the frontiers of science.

Funded by the Department of Energy Office of Science (DOE-SC), Michigan State University (MSU), and the State of Michigan, full-scale construction of FRIB began in March 2014, following six years of intensive planning and design. Located in the heart of the 5,200-acre MSU campus, the new DOE user facility—the first to be located on a university campus—will be operated by MSU and used by researchers from around the world to conduct experiments. Winning approval and funding for a nearly $1 billion facility was a major achievement for MSU and for the international teams of nuclear and accelerator physicists that designed the accelerator and made the case for it to be built at MSU.

RARE ISOTOPES

An isotope is one of the possible types of an element. Isotopes of the element sodium, for example, contain eleven protons and 7, 8, 9, 10, 11, **12**, 13 . . . 24, 26, neutrons. Only the 12 neutron isotope is seen in nature—the rest are observed only in experiments. The 25 neutron isotope has not yet been observed. A rare isotope of sodium has a very small abundance. Indeed, most are radioactive (for sodium all but the 12 neutron isotope) and have a fleeting existence with a lifetime measured in milliseconds. The information on the sodium isotopes is taken from M. Thoennessen, "Discovery of the Isotopes with $11 \leq Z \leq 19$," *At. Data and Nucl. Data Tables* 98 (2012): 933–59, which is part of the Discovery Project.

CHANGE IN USAGE: We usually mean that a rare isotope is an uncommon isotope with a particular number of neutrons, N, and a particular number of protons, Z. In past usage, this would have been called a rare nuclide.

MSU's quest for FRIB began in 1999 as what appeared to be an improbable adventure. However, over time, it became clear that MSU was a formidable competitor for the new facility. From its fledgling years in the late 1950s to the present, MSU's nuclear science program has developed steadily to become the preeminent university-based program in the United States.

■ Beginning the Journey

Early on, MSU was a minor player in the national and international nuclear science scene. But over the years, the nuclear and accelerator physicists who came to MSU, led initially by Henry Blosser, proved gifted at anticipating the most promising research opportunities that were to come and then building state-of-the-art accelerators and apparatus for their study.

First came the light-ion era, and MSU's K50 cyclotron, a device of unprecedentedly high precision, completed in 1965 and used to study nuclei using beams of protons. These high sensitivity studies pushed forward the frontiers of the field, made the Cyclotron Laboratory's first strong impact in nuclear science, and established the laboratory's reputation as a quality research organization.

Then in the mid-1970s, opportunities to determine the properties of nuclear matter using collisions of heavy ions opened a new area for study. MSU's Cyclotron Laboratory moved into this different and promising area of heavy-ion research. For this research, a new type of accelerator was needed, and MSU designed a prototype superconducting magnet as the first step toward developing a superconducting cyclotron. This was an inexpensive—and hopefully fundable—approach to developing the superconducting cyclotrons that could produce high-energy heavy ions.

In July 1974, MSU requested National Science Foundation (NSF) funding for the prototype magnet, and funds were received in June 1975. Almost simultaneously a collaboration of Midwest universities, eventually led by MSU, decided to build an ambitious regional facility involving two coupled superconducting cyclotrons. In 1976, the collaboration proposed this system to the NSF. NSF declined to fund this request—no one had shown that a superconducting cyclotron (SCC) could be built and some of the reviewers believed that it could not. NSF was willing to take a small gamble, however, and in 1977 it awarded funds just sufficient to convert MSU's prototype magnet to a working cyclotron. Construction began immediately, but later developments led to additional demands on the laboratory. These and technical difficulties with completing the cyclotron delayed K500 operation until August 1982.

In 1978, just a year following funding of the magnet conversion, the newly organized Nuclear Science Advisory Committee (NSAC)[1] scheduled a review process to determine which of nine new nuclear science facilities proposed for the U.S. should be funded. MSU slightly revised its 1976

AN AID TO NAVIGATION: ACCELERATORS AT MSU		
1965–1979	K50 cyclotron	Precise beams of protons
1977	Prototype SCC magnet	
1982–1990	K500 SCC	Heavy ions
1988–1990	K1200 SCC	High-energy heavy ions
1990–1999	K1200 SCC	Rare isotopes, high-energy heavy ions
2001–present	K500 + K1200 (CCF)	Intense beams of rare isotopes
2014–present	ReA3	Reaccelerated rare isotopes
2022	FRIB	

NOTE: SCC = Superconducting Cyclotron

SCC proposal for this review, earning the highest recommendation and Department of Energy funding. The K500 was not yet operating, but MSU now had the funds to construct a large second cyclotron, the K1200; couple it to the K500; and build an extensive laboratory to take advantage of the heavy-ion beams the coupled cyclotrons would produce. This was the Phase II project; building the K500 was now known as Phase I. The Cyclotron Laboratory had become the National Superconducting Cyclotron Laboratory (NSCL), a national user facility that would make its beams available to the international community, purely on the basis of the merit of research proposals submitted by researchers.

It was a major challenge to build simultaneously the K500, the K1200, new experimental apparatus, and a new laboratory to house it all. MSU also had to build the organizational structure of a national user facility to determine who would receive time on the cyclotrons, to give those users an opportunity to help determine the structure of the laboratory services, and to make it possible for them to use the laboratory facilities efficiently.

Challenging technological issues caused the system to be built in stages. The first stage was completed in August 1982, three years after the K50 had been closed, when the K500 cyclotron began a productive and influential program of research with heavy ions. The laboratory organized a new leadership team, developed new expertise with the use of heavy ions, and was led by a new group of researchers. The second stage was completed in 1988 when the K1200, fed from a newly constructed Electron Cyclotron Resonance (ECR) ion source, began to operate, first in a small space with limited experimental apparatus, and then in 1990 in the full Phase II laboratory.

While developing Phase II, a way had been found to use the K1200 cyclotron, followed by a fragment separator, to produce beams of rare isotopes. The fragment separator, called the A1200, was constructed inexpensively from repurposed beam line parts. It could isolate a fast-moving rare isotope from the hundreds of isotopes resulting from a collision of heavy ions and carry it to detectors for study. At that time we had no idea of how important this step would be, but eventually rare isotopes became a major area of research, using about half of all available beam time. It initiated the next movement in the NSCL's research direction—to the rare isotope era.

Operation of the Phase II facility, centered on the ECR + K1200 + A1200 accelerator system, began in 1991. We found that the use of the ECR source allowed the K1200, operating alone, to achieve much of the capability that the coupled cyclotrons had promised. There was a strong push to run the facility with high reliability and to do experiments with the newly available beams. However, there were competing priorities.

■ Looking around the Curve

Even before Phase II was operational, it became clear that an improved facility would be necessary for NSCL's long-term survival, and that a new plan was urgently needed in time for the 1996 NSAC Long Range Plan. It took some effort to find a new design that would be sufficiently powerful to provide a rich future for the laboratory and inexpensive enough to be funded.

It was not until 1994 that we had a solution in the form of the CCF, or Coupled Cyclotron Facility: ECR + K500 + K1200 + A1900 (a new fragment separator). It would, first of all, produce intense beams of rare isotopes, and the CCF's massive A1900 fragment separator would yield intensities of rare isotopes 1,000 times larger than the existing K1200 + A1200 facility. It would also accelerate heavy ions to high energies and be inexpensive enough to have a good chance of federal funding. In 1994, a proposal for a $21-million facility with a significant MSU cost share was submitted to NSF, which approved funding for the CCF.

When the CCF was completed in 2000, the earlier move toward a concentration of research on the physics of rare isotopes was completed. Beams of the rarest isotopes were extremely weak, sometimes one particle per hour or less, and NSCL researchers had to develop new techniques and models to perform experiments with these beams. Soon the NSCL became the world's most productive source of research with rare isotopes. Its research into their nuclear properties and the role they played in the evolution of stars led the world.

Although the laboratory's short-term future was now assured, we had realized, even before the completion of the CCF, that our longer-term future was again in doubt. Nuclear science is a highly competitive enterprise, and acceleratos were being developed in Japan and Europe that would soon challenge the preeminence of NSCL's CCF in the field. It was far from clear how, and if, we could meet this challenge. A new accelerator would cost around $1 billion, not more than one accelerator would be built, and no university had ever housed such an accelerator—they were always sited at existing national laboratory facilities. The NSCL, on the other hand, had arguably the highest level of expertise and experience in the production and use of rare isotope beams, which, it was widely assumed, would be the main use of the next major accelerator to be built for nuclear science.

The dominant paradigm had been to produce rare isotopes using the so-called ISOL (Isotope Separation On-Line) approach: rare isotopes were produced at rest by proton beams, which hit heavy targets, were collected, and then were reaccelerated. An ISOL task force was formed in 1998 to determine the nature of the new accelerator. At a March 1999 meeting of the task force, MSU made a presentation describing a new paradigm: rare isotopes produced by a fragmentation facility, a larger version of the CCF, would be stopped, and then reaccelerated in another accelerator to produce the low-energy beams that many researchers desired. The fast beams would also be available for experiments, providing an extremely versatile facility. In its November 1999 report, the task force adopted this paradigm and also decided that the primary accelerator would be a superconducting linear accelerator.

Following the March presentation, MSU decided to compete for this new accelerator, even while the CCF was being completed. There were major obstacles. MSU had little experience with low-energy beams and no expertise in linear accelerator technology, while the competing national laboratories did have that experience. For these reasons and the long-standing tradition of siting large new facilities at existing national laboratories, few would have predicted that MSU would be chosen to build FRIB. Indeed, the opinion of many faculty members at MSU was that MSU's chance of getting FRIB was, at best, one in ten.

Nevertheless, the NSCL, facing an even dimmer alternative, decided to press ahead. To have a chance of competing, MSU would have to make a convincing case that it could construct a high intensity linear accelerator, that fast beams could be stopped efficiently and reaccelerated, and that fast beam experiments were important.

■ Reinvention: Continuity and Change

Thus was launched a grueling—and hectic—journey through many review panels, redefinition upon redefinition of what the accelerator should be, writing and rewriting of proposals up to the time the final proposal was ultimately presented, and meetings, meetings, and more meetings with representatives of every group that could be strategically influential in the DOE's ultimate decision.

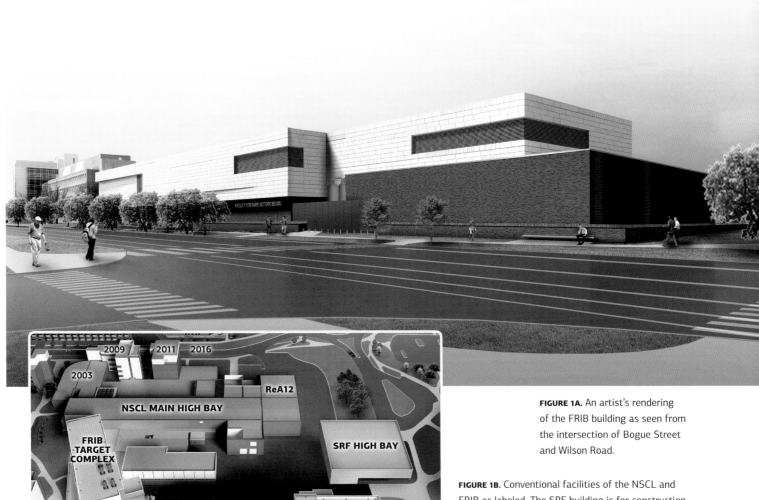

FIGURE 1A. An artist's rendering of the FRIB building as seen from the intersection of Bogue Street and Wilson Road.

FIGURE 1B. Conventional facilities of the NSCL and FRIB as labeled. The SRF building is for construction of superconducting radio frequency systems. The buildings with construction dates at the top of the figure are mainly offices; another office building will be completed in 2016. The FRIB accelerator and experimental areas are toward the bottom. A more detailed view is shown in figure 1c.

FIGURE 1C. The accelerator and experimental systems and experimental area locations inside the Facility for Rare Isotope Beams.

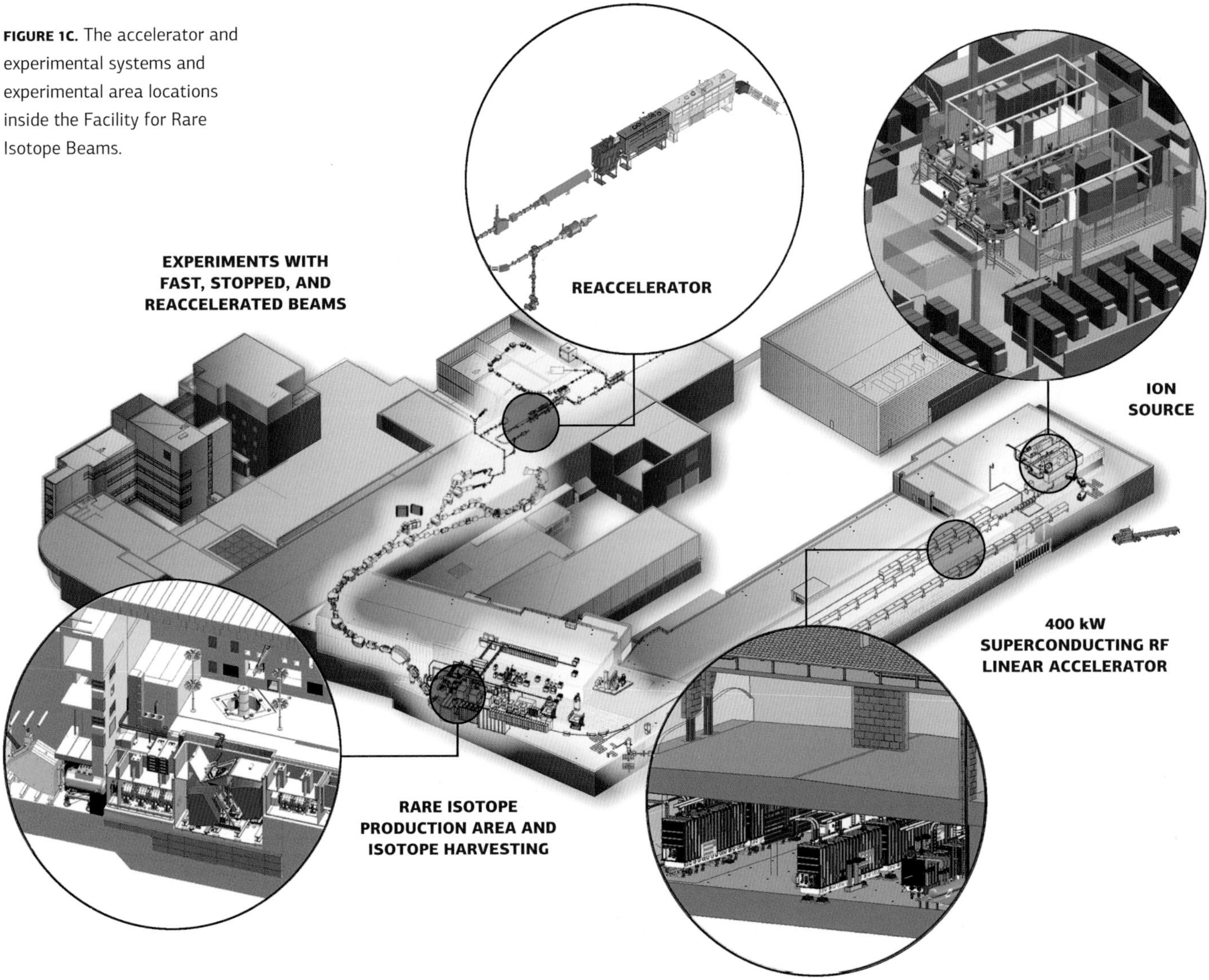

EXPERIMENTS WITH
FAST, STOPPED, AND
REACCELERATED BEAMS

REACCELERATOR

ION
SOURCE

400 kW
SUPERCONDUCTING RF
LINEAR ACCELERATOR

RARE ISOTOPE
PRODUCTION AREA AND
ISOTOPE HARVESTING

Perhaps decisive for NSCL's position to press on in the competition was the realization, mentioned previously, that low-energy beams could be produced efficiently by slowing fast beams in gas-filled devices and then reaccelerating them. This had the strong advantage that the process was fast, so isotopes with shorter lifetimes could be produced, and it was independent of chemical effects that made it difficult for ISOL to produce beams of some elements. Thus, one accelerator system could efficiently produce both fast and slow beams of rare isotopes. Experiments with fast beams could be done with upgrades of the apparatus already developed at the NSCL, while the use of low-energy beams would require the development of new low-energy accelerators. MSU argued persuasively for this new concept, and on December 11, 2008, it finally was selected for the design and establishment of FRIB. The accelerator complex that will comprise FRIB is shown in figures 1A, 1B, and 1C.

During this selection process, in September 2006, MSU and NSCL decided to build a reacceleration facility, funded by MSU, as part of its eventual contribution to FRIB, should MSU win it. If FRIB was not built, or was not located at MSU, it would also initiate a new direction for research at the NSCL. The new facility, ReA3 (for Reacceleration of ions to 3 MeV/nucleon), would use fast nuclei from the A1900 fragment separator, slow them in a gas or cyclotron stopper, and then reaccelerate them in a superconducting linear accelerator to energies of 3 MeV/nucleon or more, depending on ion mass. It would provide unique radioactive beams, especially suited for studies of reactions of interest in astrophysics. A gas stopper was built for the NSCL at Argonne National Laboratory (ANL) and is operational. An inverse-cyclotron-based beam stopper, a massive device, is, in mid-2015, in the final stages of construction at NSCL.

Now that FRIB is being built at MSU and ReA3 is operational, ReA3 provides a demonstration of the reacceleration of stopped beams that is at the heart of the MSU approach to FRIB. ReA3 also served as a test bed for construction of the superconducting cavities that will be a major part of FRIB and serve as part of a reaccelerator for FRIB. This allowed the problems inevitably encountered with new superconducting cavities to be solved in a small-scale environment that was much more forgiving than the FRIB project would be.

The years 2009 to 2013 at the new FRIB project were spent refining costs and developing detailed construction plans that, through a series of DOE reviews, finally convinced DOE-SC that construction could begin. A groundbreaking was held in March 2014, and in July, the first concrete was poured in the enormous tunnel that will house the FRIB accelerator. On August 26, 2014, DOE-SC issued

Critical Decision CD-3b: Start of Construction of the Accelerator and Experimental Systems for the FRIB Project. CD4 will be the commissioning into service, currently planned for 2022.

During the next seven years, the laboratory will be a beehive of activity as civil construction is completed and the many technical components of FRIB are assembled and installed: the ion source front end, driver linac, beam delivery system, isotope production target, radiation shielding,[2] fragment separator, experimental areas, and all their many subcomponents. Undoubtedly, there will be difficulties to overcome and technical challenges to be met. But when FRIB is complete in 2022, or perhaps somewhat earlier, the MSU Cyclotron Laboratory will complete its largest move: a leap into the FRIB era.

In the following pages, I detail the sequence of events that led the MSU Cyclotron Laboratory up from nothing in 1958 to FRIB, and attempt to delineate the characteristics of the laboratory that led to that success.

The Opportunity and the Will

In the early years following its founding in the late 1950s, the MSU Cyclotron Laboratory played an insignificant role in the U.S. nuclear science program. Since then it has grown into a major facility and is home to the highest-ranked university nuclear physics program in the U.S. In 2008, MSU was chosen by the DOE to build FRIB, a forefront international user facility for nuclear science that will investigate the properties of rare and extremely neutron- or proton-rich isotopes and their role in the cosmos.

Two circumstances foreshadowed MSU's extraordinary trajectory in nuclear science. First, by the end of World War II, nuclear science had emerged as an important science frontier, but the tools available for studying nuclei were inadequate to the task at hand. More capable accelerators, producing higher energy and more precise particle beams, were required. By serendipity, the right visionary and the right idea converged on the Michigan State University campus, and the stage was set for exceptional progress.

Second, from 1941 to 1969, Michigan State University[3] was headed by John Hannah, a dynamic president who was determined to[4] ". . . mobilize the faculty, the alumni, the friends of the college, the people of the state who supported it, the students and their parents, and to try together to make Michigan State a distinguished university."

MSU had joined the Big Ten Conference in 1950 but was not yet competitive in research and academics with its Big Ten peers. Hannah believed that the quality of a university depended on the quality of its faculty, and that one could only attract quality faculty with forefront facilities and the promise of distinguished academic and research programs.

■ Beginnings of Nuclear Science

Nuclei are usually studied by observing what particles are emitted when a nuclear projectile hits a nuclear target. A common analogy compares this to learning the workings of a wristwatch by observing what gears and jewels are emitted when the watch is thrown against a wall. Since nuclei are so small, the process is still more difficult because one cannot actually see the nature of the emitted nuclear "gears" and must interpret the observations using a theoretical model of the process.

In a pioneering experiment, see figure 2, Nobel Prize–winning physicist Ernest Rutherford and his collaborators bombarded a gold target with helium nuclei (a.k.a. alpha particles) from a radioactive source. When they observed that some of the helium nuclei were scattered backwards, Rutherford (1911)[5] concluded that the only model of a gold atom that reproduced this behavior was one in which almost all the mass of the atom was concentrated in a tiny nucleus, much smaller than the atom. This discovery of the nucleus marked the beginning of nuclear science.

We will often speak of a particle beam, a pencil-like array of particles moving in nearly the same direction. In the Rutherford experiment, an alpha particle source that produced a weak beam of particles of a single type was adequate. But for a more detailed description of a nucleus, one needs a variety of nuclear probes with variable and well-defined speed and direction. As a result, the history of nuclear science is embodied in the accelerators it uses to produce precise nuclear beams with speeds high enough to overcome the electric (Coulomb) repulsion that prevents two positively charged particles from closely interacting.

By about 1930, electrostatic accelerators were being used for this purpose. In the most common electrostatic accelerator, the Van de Graaff, a belt, like that used to transfer power in factories prior to the invention of the electric motor, carries electric charge to an electrode (often called a terminal) to increase its voltage. Ions, charged particles, gained energy by being produced on a terminal held at high voltage, thousands to millions of volts, and let fall (accelerated) to a grounded, zero voltage

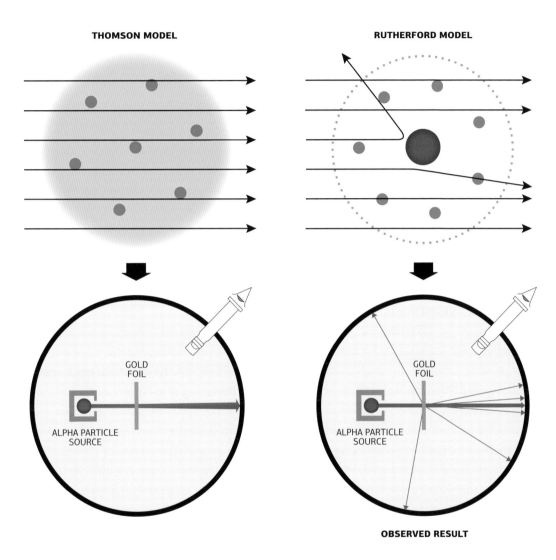

FIGURE 2. In the Geiger Marsden experiment (also called the Rutherford gold foil experiment) shown schematically here, a collimated beam of alpha particles from the radioactive decay of radon passes through a gold foil. In the earlier Thomson Model (plum pudding model) of the gold atom, positive and negative charges were assumed to be uniformly distributed throughout the atom and only small angular deviations were expected. But occasional large deviations were observed leading Rutherford to propose a model in which most of the atom's mass is located in a small nucleus.

FIGURE 3. Ernest O. Lawrence came to UC-Berkeley in 1928, having received his PhD at Yale. By 1930, he was working on his first cyclotron. He built a sequence of cyclotrons, of increasing size and energy between 1931 and 1942, beginning with a five-inch-diameter device and culminating with a 184-inch device that was later used as a synchrocyclotron.

electrode. Speeds, or velocities, of a few percent of light's velocity could be obtained, and the particle beams could be very precisely controlled. However, their energy was limited by the voltage one could achieve on an electrode. Voltages above five million volts are hard to achieve and are expensive, and electrical sparks are common and often damaging. The record of 25.5 million volts was achieved at a Tandem[6] Van de Graaff[7] accelerator located at Oak Ridge National Laboratory (ORNL); most Van de Graaffs reach much lower voltages.[8]

For what follows, it's necessary to understand how energy is specified. In nuclear physics, energies are usually given in electron-volts or eV (Multiples of eV, millions or billions of eV, denoted as MeV and GeV, are in common use). An eV is the energy a proton would have if it were emitted from the positive terminal of a 1 volt battery and allowed to accelerate to a zero voltage electrode. If a particle is a nucleus, with several nucleons (neutrons and protons), the energy is given as MeV per nucleon or MeV/nucleon.

■ Why Cyclotrons?

The drive to higher velocities relied on being able to apply a voltage repetitively, rather than a single time, as in the electrostatic accelerators. A device that could do this, the cyclotron, was invented[9] by Ernest Lawrence (see fig. 3) in 1930. In a Lawrence cyclotron, a positive ion is injected into a magnetic field and moves in a circular orbit perpendicular to the direction of the magnetic field and with an orbital time that does not depend on the ion's speed. During most of the orbit it is inside a "D" shaped metal electrode (a dee), feels no electric voltage, and just coasts. But if the voltage on the dees is changing at the orbital frequency, when the ion leaves one dee, the other dee will be at a negative voltage, and will pull the positive ion toward it and accelerate it (see figs. 4 and 5).

Thus the particle will be accelerated on each orbit, will gain energy and speed, and spiral outward. As the energy gets high enough, however, relativistic effects cause the ion mass to increase. This causes the ion's orbital frequency to decrease, it falls out of synchronization with the frequency of the accelerating voltage, and acceleration ceases. This limits the energy of a Lawrence cyclotron to around 20 MeV for protons.[10] But higher energies were desirable, as noted in figure 6.

One obvious solution to this problem is to change the radio frequency as the ion spirals outward to stay in synchronization with the orbital frequency. Such frequency-modulated (FM) cyclotrons

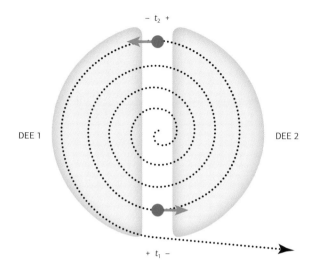

FIGURE 4. How a cyclotron works: a schematic view of the cyclotron acceleration process. A red proton's position is shown at two times, when it is in the gap between the gray-shaded metal dees. (While it is inside the dees, it feels no electrical forces.) At time t_1, the proton is at six o'clock, the voltage on Dee 1 is positive and that on Dee 2 is negative, so the proton sees an electric force in the direction of the arrow and is accelerated (speeded up). Half an RF cycle later it is at twelve o'clock, the Dee voltages have reversed, and the particle sees the force shown and is again accelerated. So, it moves in an expanding spiral accelerated twice by the RF voltage on each turn.

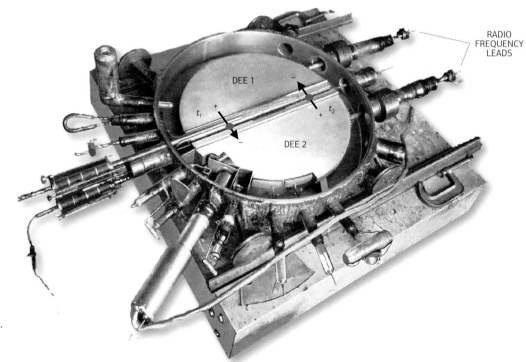

FIGURE 5. Here we see an actual Lawrence twenty-seven-inch cyclotron built in 1932. The two D-shaped dees are labeled as are the conducting leads at two o'clock that bring the radio frequency voltages to the dees. The entire device is located in a magnetic field whose direction is perpendicular to the plane of the dees, so a particle with a constant energy would move in a circle. When it is inside the conducting metal of the dees, and not at the gap between them it feels no electrical force. It is accelerated when it crosses the gap between the dees. The description is essentially that of figure 4 except that acceleration occurs at two and eight o'clock.

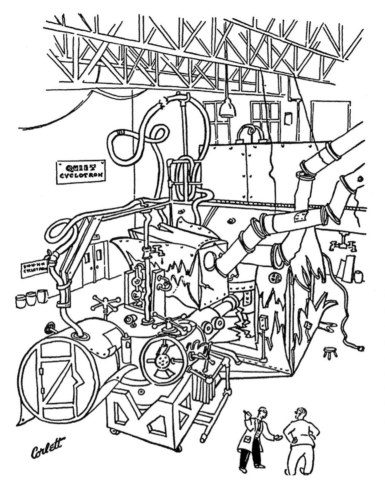

FIGURE 6. "Toughest damn atom I ever saw!" (from the California *Pelican.*) The drive for higher energies that led to synchrocyclotrons and sector-focused cyclotrons is epitomized in this cartoon. One needed higher energies to overcome the Coulomb barrier for heavy nuclei: for alpha particles bombarding lead the barrier is over 30 MeV. In addition, certain reactions will produce convincing results only at high energies.

or synchrocyclotrons were built and worked; some achieve energies one hundred times as high as a Lawrence cyclotron. The 160 MeV FM cyclotron at the Harvard University Cyclotron Laboratory, for example, was completed in 1949 and operated until 2002.[11] Unfortunately this approach has disadvantages: the beams are pulsed and are not sufficiently intense or precise for many studies in nuclear science.

For these reasons another approach became far more common. It is to keep the rotation time constant, in spite of the relativistic mass increase. Doing this introduces its own complications: the magnetic fields must vary in a complex way as the ion traverses a nearly, but not quite, circular orbit in the field. These devices are called isochronous or sector-focused cyclotrons. To design such a cyclotron is difficult and requires computational facilities that were not available until the mid-1950s, and even then had limited capability. As a result, the first of these cyclotrons were crude instruments, but could reach higher energies than a Lawrence cyclotron and higher intensities than an FM cyclotron and so were more suitable for studies of nuclei.

The influence and dominance of the sector-focused cyclotrons can be seen in compilations of existing cyclotrons presented in the conference proceedings of *Cyclotrons-1972*[12] and *Cyclotrons and Their Applications, 1978*.[13] Of the seventeen FM cyclotrons listed there, only two were built after 1964. By 1979, only twelve FM cyclotrons were operating, and no new FM cyclotrons had been built. Prior to 1965, there were twelve sector-focused cyclotrons worldwide, and none of these were precision devices. But, by 1972, just eleven years after the first sector-focused cyclotron was built, there were seventy such cyclotrons, many with better precision and intensity than the FM cyclotrons.

The first cyclotron built at MSU—the K50—was a high-precision, sector-focused cyclotron. It was completed in 1965 and established a tradition of excellence at MSU, giving the institution the credibility to propose and construct a series of more and more powerful accelerators. In this sense it is responsible for the present world-leading status of the MSU Cyclotron Laboratory.

THE EVOLUTION OF MICHIGAN STATE UNIVERSITY

The first century at MSU is discussed in detail by Madison Kuhn in a book published on MSU's 100th Anniversary in 1955.[14] Fifty years later, a history of MSU's College of Natural Science was prepared by Clarence Suelter on MSU's 150th anniversary.[15] Material from these sources is the basis for this section.

Once Michigan achieved statehood in 1837, the first legislative session established a state university (known as "The University") at Ann Arbor and charged that: ". . . there shall be a Department of Agriculture, with competent instructors in the theory of agriculture, including vegetable physiology and agricultural chemistry, and experimental and practical farming and agriculture."[16] However, due to lack of funds, or because the University's orientation was toward a classical education with little emphasis on science, it did not address this charge. Nor did the Michigan State Normal College (now Eastern Michigan University), founded in 1853, do so. But, Michigan continued to suffer from poor crop yields and quality, and a growing faith in scientific agriculture suggested that the state needed better educated farmers.

In 1849, the Michigan State Agricultural Society was founded, and it was determined to obtain public support to establish an Agricultural College. This goal was delayed by competition between The University and the Normal College, each wishing to preserve its academic primacy. It became apparent that agriculture was not a major interest of these institutions, and although they made token attempts, the courses were taught neither well or efficiently. By 1854, the general opinion among agricultural leaders reached a consensus and the Society passed a resolution stating: ". . . that an Agricultural College should be separate from any other institution."[17]

Finally, on February 12, 1855, a law was signed establishing The Agricultural College of the State of Michigan. It was "directed to improve and teach the science and practice of agriculture." It was charged to teach an "English and Scientific Course." Such a course "normally included an introduction to various sciences and extensive study of the English language and literature, mathematics, philosophy, geography, political economy, and history." To indicate that the College must teach more than the typical amount of science, the act listed "Natural Philosophy, Chemistry, Botany, Animal and Vegetable Anatomy and Physiology, Geology, Mineralogy, Entomology, Veterinary Art, Mensuration, Leveling and Political Economy, with Bookkeeping and the Mechanical Arts which are directly connected with agriculture."[18]

The College was to occupy an experimental farm and be within ten miles of Lansing. Manual labor was to be required of all students. It was some years before the ambitious goals laid out above could really be met. In 1860, only four professors were available to teach all the scientific courses.

In 1862, the Morrill Act,[19] sometimes known as the land grant act, granted to each state 30,000 acres of public land for each Senator and Representative under apportionment based on the 1860 census. "Proceeds from the sale of these lands were to be invested in a perpetual endowment fund to provide support for agricultural and mechanical arts colleges in each of the states. The legislation

THE NAMES OF MICHIGAN STATE UNIVERSITY

Michigan State University (MSU) has had various names at various times in its evolution as noted below. In most of this text it is referred to as MSU.

1855	Agricultural College of the State of Michigan
1861	State Agricultural College
1909	Michigan Agricultural College
1925	Michigan State College of Agriculture and Applied Science
1955	Michigan State University of Agriculture and Applied Science
1964	Michigan State University

provided support primarily for educational purposes; supporting experimental work was not given a high priority."[20] John Hannah later noted the impact of this act:

> In the new land grant institutions, there was to be no denying or downgrading of learning, but the emphasis was to be on the *utilization* of learning for the service of people. And for the first time in history, it became respectable to emphasize college courses that would produce more effective farmers, engineers, housewives, or business people, or that would help them earn a living and develop personal philosophies that would make life more satisfying than it might otherwise have been.[21]

Initially the number of students at MSU was small, with typically fewer than 100 on campus before 1870 and fewer than 500 until 1898. However, growth then accelerated; over 8,000 students were on campus in 1941 when John A. Hannah (see fig. 7) became president of MSU.

The years following World War II brought remarkable changes to MSU. Enrollment doubled between the 1944–45 and 1945–46 academic years, reached 25,000 by 1956 when the quest for nuclear science began, and 49,000 in 2012–13. Of course, many new people had to be hired to teach these students, and the growth of scale made it possible to consider investments in improving the university, especially its faculty. When MSU joined the Big Ten in 1950, it was clear that, academically, MSU lagged far behind its peers. President Hannah's strong motivation to make MSU a distinguished university had positive implications for the growth of research in the Physics Department and in other Science Departments.

A measure of the success of Hannah's initiative over the long term is reflected in recent changes of the name and self-image of the institution. MSU is now among the top one hundred universities worldwide.[22]

FIGURE 7. John Hannah, President of MSU from 1941 until his retirement in 1969.

The MSU Physics Strategy, 1955–62

Physics was barely mentioned in Madison Kuhn's 1955 history of MSU's first century, perhaps because research played a minor role in the Physics Department. The first Physics PhD was awarded in 1935, the second in 1945, and only eleven in the next ten years. However, in 1954, the department undertook a long-range planning exercise,[23] and the resulting plan set a goal of developing a reputation in research and teaching, with future emphases in solid state and nuclear physics.

The department grew rapidly and by 1959 had a significant physics research effort[24] in a number of research areas, carried out by five professors, five associate professors, fourteen assistant professors, and twenty PhD students. In the previous three years thirteen PhD and seventeen MS degrees had been awarded.

The general advancement of physics was strongly supported by the MSU administration. Milton Muelder, then-Dean of Science and Arts, stated in 1978[25] that his general attitude was "to support physics, chemistry, and math as fast and strongly as circumstances and opportunity warranted." This is corroborated[26] by Richard Schlegel, acting Head of Physics from December 1955 to September 1956, who stated that "Muelder told him that there was an administrative interest in developing as strong a Physics Department as possible." During his short tenure Schlegel added three faculty members to the department.

FIGURE 8. Joseph Ballam came to MSU in the summer of 1956, after receiving a PhD from UC-Berkeley, and serving as an Assistant Professor at Princeton University. He chaired the Ballam Committee that developed plans for a future accelerator at MSU.

However, the development of the MSU Cyclotron Laboratory owes its existence much more directly to President John Hannah's interest in developing major initiatives that would bring credit and reputation to MSU. When Hannah heard that fourteen universities had formed the Midwestern Universities Research Association (MURA) in September 1954,[27] MSU applied for MURA membership.[28] Since MURA was formed to recommend a design for a high-energy accelerator to be built in the Midwest,[29] and no one at MSU was involved in related research, this was a future-oriented gamble.

MSU became a MURA member in 1956, paying a membership fee of $15,000 (equivalent to $130,000 in 2013 dollars), and shortly thereafter the Physics Department received a new position with the stipulation that it be filled by someone who could participate in MURA activities. Princeton physicist Joseph Ballam[30] (see fig. 8) was attracted to MSU in the summer of 1956, and while at MSU continued his experimental work at Brookhaven National Laboratory. His experience with "Big Science" played an important role in the founding of the Cyclotron Laboratory.

■ Choosing What to Build

Schlegel states that there was apparent administrative interest in bringing laboratory nuclear or meson physics to MSU. A Nuclear Physics Planning Committee, later called "The Ballam Committee," (BC) was appointed in the fall of 1956 by the next acting department chair, Robert Spence, to study various nuclear accelerators and propose one that would be appropriate for MSU. Guidance to the Committee from the Dean's office[31] was that the department should "not imitate any other institution and . . . select an energy range and type of activity which would make a contribution; hence gain distinction to the Physics Department and MSU." The BC was chaired by Joseph Ballam; other members were D. J. Montgomery, C. D. Hause, W. Kelly, G. Beard, and J. Kovacs. Kelly and Beard were nuclear physicists and Kovacs was a newly appointed elementary particle physics theorist. The committee considered possible devices that gave energies from several eV to 25 GeV, involving five different machines of all sizes, from electrostatic particle accelerators and nuclear reactors to large proton and electron synchrotrons.

During a meeting on February 22, 1957, the BC narrowed these choices to three:

- a 500 MeV cyclotron, a scaled-down version of a 750 MeV machine proposed by Oak Ridge National Laboratory (ORNL);
- a 10 MeV Van de Graaff that was commercially available; or
- a 40 MeV heavy-ion cyclotron to accelerate carbon, nitrogen, and oxygen beams.

By the time the BC gave its preliminary assessment to the Physics Department in meetings on March 5 and 8, 1957,[32] it had concluded that the ORNL copy was out of the department's reach.

The first question addressed in the meetings was whether to go ahead at all, given the concern that funding for other departmental research might be negatively affected. Then, a motion was made and approved unanimously that the department promote one of the two machines put forward by the BC. The issue of competition with other departmental research came up again in the second meeting, and after additional discussion resulted in the passage of a second motion: that the department should request a separate, distinct budget for the accelerator.

A second question was: "which machine should be built?" Of the two remaining possibilities, the BC favored the heavy-ion cyclotron, arguing first, that those who built the machine would have an intimate knowledge of its properties, leading to easier maintenance and more imaginative research; and second, that the experience gained would augment the department's strength and heighten its ability to progress rapidly in nuclear physics. The BC's rough estimate of the heavy-ion cyclotron's cost was $1,300,000, including the cost of a building.

On a unanimous vote, the Physics Department accepted the BC's recommendation and estimate. Ballam, however, felt that a better cost estimate could be gained during a one-week visit by the committee members to ORNL. A motion was made and passed that the department chair should approach the administration with a request for planning money. This request was presented to a meeting of President John A. Hannah; Vice President for Academic Affairs Thomas H. Hamilton; Dean of Science and Arts Milton E. Muelder; Dean of the Graduate School Thomas H. Osgood; and Physics Professors Robert Spence and Joseph Ballam. Together they decided that a small amount of money—around $2,000—would be granted to the BC to complete the study of the heavy-ion accelerator, especially its projected cost, necessary personnel, and design and construction time.

After working on a remarkably fast time scale, in June 1957 the BC presented its report[33] to the Physics Department. In twenty-seven pages this report outlined:

- the scientific motivations, pointing out in particular that the field of heavy-ion physics was wide open so that MSU could compete more easily with established institutions;
- a recommendation for the construction of a heavy-ion cyclotron, based as much as possible on ORNL designs; and
- a moderately detailed conceptual design that estimated individual system costs.

The total cost was $1,037,390, not including a building. The Report requested $95,500 from the university, primarily for personnel, to produce a proposal-ready design for the cyclotron by late spring 1958. The department approved the report with only two negative votes.[34]

The BC report outlined two characteristics of the future MSU nuclear science program: a preference for building equipment, rather than buying it off the shelf, and an expectation of strong MSU support for the program. The build-it-yourself approach was daring because the required infrastructure to do so was lacking. Success would have to depend on strong entrepreneurial characteristics of the future project leader.

■ Selecting a Leader

Thus, the immediate business at hand was to find a project leader, and also to find someone to head the Physics Department. No one in the Physics Department wanted to be Head, and the search was long and difficult.

When Sherwood K. Haynes from Vanderbilt finally applied to head the department, Dean Muelder jumped on the opportunity and tendered him an offer, giving the Physics Department little choice.[35] Haynes immediately began looking for someone to head what had become known as the Cyclotron Project.

He first approached two relatively senior people, Bernie (B. L.) Cohen from Pittsburgh and Alex Zucker from ORNL. When they refused, he turned to a younger man, Henry Blosser, then also at ORNL working on its cyclotron project. Blosser recalled that:

a couple of guys from the (MSU) Physics Department, Bill Kelly and George Beard, came to Oak Ridge looking for people to head up a cyclotron project which was being dreamed about. I had lunch with them one day, and then a bit later, a department head-designate, Sherwood Haynes [who] had accepted a position to come to MSU, [heard] a presentation at an information meeting about the cyclotron we had built there in Oak Ridge. He came up afterwards and asked would I be interested in a position . . .[36]

Blosser interviewed, and letters of reference were obtained from Cohen and from Robert Livingston, director of the Electronuclear Research Division at Oak Ridge. These letters stressed Blosser's technical strengths, but also that he tended to be "very demanding of his own way" and "very unhappy when he is working on a projects in which he is not in full charge."[37]

An offer of an Associate Professorship at MSU at an annual salary of $12,000 was made to Blosser on January 21, 1958,[38] and he promptly accepted the offer (see fig. 9). The official appointment was to begin September 1, 1958, but it was later moved up to June 1. An interim half-time appointment began on March 1 to get the project started, and MSU committed $150,000[39] to support the project until federal funding was obtained.

FIGURE 9. Henry Blosser, Professor of Physics and founding director of the Cyclotron Laboratory at MSU.

■ Designing and Funding the Cyclotron

Blosser immediately began work on a proposal for a heavy-ion cyclotron, consistent with the proposal of the Ballam Committee, but with new accelerator design ideas that had been under development by Blosser and others at ORNL. Their emphasis was on delivering uniquely precise beams from the cyclotron. Progress was remarkable. By May 1958, the main features of the magnetic field had been mocked up in tests at ORNL. A model magnet facility at MSU was scheduled for completion by June 15, 1958. The MSU Electrical Engineering Department undertook the design of the radio frequency system; mechanical design and engineering issues were contracted to Brobeck and Associates in Oakland, California; and building design by MSU architects was underway.

However, it remained for MSU to make a credible case for federal funding. Proposal reviewers had to be convinced that, if the cyclotron was funded, the people at MSU could build it. Blosser's expertise was in accelerator design, and it was not yet clear that he had the management and construction skills that would be needed to convert a paper design into a working accelerator. Reviewers also had to be convinced that important science could be done with the accelerator. Unfortunately, the nuclear physicists then at MSU, William Kelly and George Beard, had interests very different from most of the research that would be done with the new cyclotron and could not make a credible case for it.

Blosser approached this problem in the direct way that those who worked with him learned to think of as typical. He assumed that any ambitious person would be interested in working at the new MSU Cyclotron Laboratory, and in late May 1958, he contacted D. Alan Bromley, then in a senior position at Chalk River Nuclear Laboratories in Ontario, Canada. Bromley expressed strong interest in an MSU position, and Blosser persuaded the MSU Physics Department to offer him a Full Professorship, with principal responsibility for the experimental program at the new cyclotron.[40] The position was to begin in September 1959. By fall 1958, however, it had become clear that the new cyclotron could not run before June 1962, and Bromley requested a delay in his decision.[41] However, he coupled this request with an offer to work on the nuclear physics part of the upcoming MSU proposal. He later wrote a substantial part of the 1958 proposal to the Atomic Energy Commission (AEC).

The MSU response was unusual, clearly reflecting Blosser's strong positive evaluation of the influence of a senior nuclear physicist like Bromley, both on funding of the proposal and on the future of the experimental program. MSU agreed to hold open the offer until the Physics Department decided to close it and not to make another offer in the position prior to Bromley's final refusal.[42] This process ended on November 19, 1959, when Bromley informed Blosser and Haynes that he had received offers from Yale and Florida State and, after a long period of stalling, had decided to accept the Yale offer, the position to begin February 1, 1960. Nonetheless, Bromley remained an influential friend of the MSU Cyclotron Laboratory. He later had an extremely distinguished career at Yale, had great influence in the development of nuclear science, was Director of Yale's A. W. Wright Nuclear Structure Lab, and served as U.S. President G. H. W. Bush's Science Advisor.

On December 11, 1958, MSU submitted a proposal to the AEC,[43] a 133-page document for a "Nuclear Research Facility in the Medium Energy (50 MeV) Range, Utilizing a 64" Variable Energy Multi-particle Cyclotron." It could produce energies ranging from 20 to 40 MeV for protons and 28 to 55 MeV for carbon ions. The proposed experimental program emphasized studies that would take advantage of the unique characteristics of the proposed MSU cyclotron: energy variability, precise beams, and high current.

The choice of energy was comfortably above that accessible to Van de Graaff accelerators and below that of the much less precise cyclotrons already operating. It was, hopefully, in a "sweet spot" where much interesting physics could be done using its unique properties, yet the cost made federal funding feasible, if probably difficult, for a new laboratory. The ten proposers were Blosser, Haynes, Beard, and Kelly, plus six Physics Department faculty, active researchers but with little interest in nuclear physics.

At this point the proposed site of the new Cyclotron Laboratory was just south of the Physics-Mathematics Building (now housing the Psychology Department) on MSU's North Campus. The present site on Shaw Lane in mid-campus was decided upon later. The proposal requested $1,473,900 from the AEC with an additional $730,300 contribution from MSU, mostly for construction of the laboratory building.

ANTICIPATED ADVANTAGES OF A SECTOR-FOCUSED CYCLOTRON

A digression on high-energy cyclotrons is necessary to appreciate what MSU had accomplished. The classical Lawrence cyclotron could not accelerate protons to energies much above 20 MeV.[44] It relied on giving protons, moving in a circular orbit in a magnetic field, small electrical kicks during each orbit. But the relativistic increase in mass at higher energies caused the orbital time to fall out of synchronization with the radio frequency voltage that supplied these kicks. To restore the synchronization, the magnetic field would have to increase as the orbit moved out in radius, but this would defocus the beam and the beam particles would strike the interior parts of the magnet rather than remaining in well-defined orbits. In order to restore focusing, it was necessary to introduce a magnetic field that varied along the (almost) circular orbits. There would be strong and weak magnetic fields following each other like slices in a pie. This led to the name: sector-focused cyclotron.

Computation of the orbits in such a field was difficult, indeed impossible in practice until the advent of fast digital computers. The techniques for doing so were developed at MURA in the mid-1950s. They were first applied to an actual cyclotron by Blosser and his collaborators at ORNL and published in 1958.[45] A critical focus of these studies was the means to extract the beam from the cyclotron and send it down a beam pipe toward experiments. It appeared that one could extract beam from a single orbit, and that if one could do so, a beam of unprecedented precision could be obtained. The December 1958 proposal described the results of these techniques in detail.

The bending power of a cyclotron is often expressed by a parameter K, approximately the energy in

FIGURE 10. A model magnet showing the sectored structure. It is a one-sixth scale version of the proposed sixty-four-inch-diameter cyclotron magnet. In this picture the top half of the magnet has been removed to better show magnet structure. The magnet has three raised iron poles where the field will be high and three valleys where the field will be low, forming three magnetic sectors.

MeV of the protons that can be accelerated. More generally, for an ion of charge Q and mass number A, the energy per nucleon is given by $K(Q/A)^2$. It is common to refer to a cyclotron with K = 50 as a "K50 cyclotron" or as a "K50." That notation is used here for brevity and for consistency with the later MSU cyclotrons. However, during its lifetime, the K50 was generally referred to as the MSU Cyclotron.

MSU made several important hires at about this time, including Morton (Mort) M. Gordon.[46] Blosser was fond of remarking that this was the only day on which the staff of the new Cyclotron

Laboratory doubled. Gordon came from the University of Florida, but had worked with T. Welton at ORNL and had written the definitive papers on resonant extraction from sector-focused cyclotrons, a technique that would be further developed at MSU and would form the basis of the precision beams from the new cyclotron. Gordon was a remarkable individual who continued his prolific theoretical research and teaching although he had become blind. Thelma Arnette came from ORNL and specialized in computer programming; this would be her emphasis at MSU. Hugh McManus was a theoretical physicist from the UK who had been working at the Chalk River Nuclear Laboratory in Deep River, Ontario.

Hiring quality staff has been a concern from the beginning of the laboratory. Initially the task was to find skilled accelerator physicists, which, even for a relatively small project like the K50, was difficult for an unknown, unfunded facility. Recruiting a nuclear research faculty was a complicated and longer term effort. Most faculty were nuclear physicists and held positions in the Department of Physics. A smaller number were nuclear chemists and had positions in the Department of Chemistry. The laboratory made successful early hires, but continuing this success, as the laboratory grew, required continuing close attention.

■ The Funding Saga

Nothing about the fate of the proposal was heard from the AEC until September 1959, when MSU was notified that it had received high technical scores from the AEC Division of Research, but that it and all related proposals had been rejected for Fiscal Year (FY) 1961 funding because of a constrained federal budget.[47] In a letter to President Hannah dated November 2, 1959, Muelder reported the results of a visit to AEC.[48] The Acting Director of AEC's Research Division had said that if MSU was also turned down by AEC in the next budget cycle, FY-1962, it should look elsewhere for funding because a large number of commitments would have built up during of the absence of accelerator funding over the previous two years. In addition, new projects at Stanford (SLAC), Argonne National Lab (a linear accelerator), and MURA would be competing for funds.

The crucial extraction studies upon which the precision of cyclotron beams depended, and upon which the 1958 proposal had been based, were mainly done using ORNL computers. Late in 1959, however, the AEC expressed willingness to consider[49] a proposal for studies related to beam extraction

FIGURE 11. Lawrence Von Tersch (*standing*) and (Martin) Glen Keeney examining parts of the Michigan State Integral Computer (MISTIC), built in 1957 on the fifth floor of the MSU Computer Center. It was modeled closely after the University of Illinois ILLIAC computer and took 100 (1,000) microseconds for addition (multiplication) operations. The central processor was ten feet by eleven feet by two feet and contained 2,610 vacuum tubes. Input was by teletype, paper tape, or punch-cards.

from sector-focused cyclotrons in general, but not specifically the proposed MSU machine. Blosser and Gordon submitted a proposal to AEC for studies of "Resonant Extraction in Sector-Focused Cyclotrons" on November 4, 1959. Funding of $165,000 was obtained for these studies in two annual installments beginning February 15, 1960. This was the largest federal award for research in the Physics Department, and perhaps the university, at that time, and it lent on-campus credibility to the work of the cyclotron group.

These new studies used the MSU MISTIC computer (see fig. 11).[50] MISTIC (for Michigan State Integral Computer) was a state-of-the-art computer and one of a limited array of technical strengths at MSU. It had recently been upgraded with additional magnetic core memory and had an operating system similar to the ORNL computer, making transfer of programs straightforward.

The research (see fig. 12) also took advantage of MSU's new 1/6 scale model magnet to provide the relevant magnetic fields. Its results provided a more detailed understanding of resonant extraction and strengthened the second proposal to the AEC in June 1960 and a similar proposal to NSF in December 1960.

In February 1960, the AEC requested further copies of the proposal for review, apparently for the FY-1961 or 1962 budget. In March 1960, the results of an AEC review of about ten accelerator

FIGURE 12. Henry Blosser shown making high-precision measurements of the magnetic fields produced by the model magnet shown in figure 10. The model magnet is shown on the right and a repurposed milling machine table is shown holding the device that measured the magnetic field.

FIGURE 13. "The Cyclotron as Seen by the Theoretical Physicist." The capability of the proposed cyclotron was a great advance over past cyclotrons. This was made possible by a deep understanding of orbit dynamics by Blosser and Gordon, coupled with the computational power of the ORNL computers and of MISTIC. It is not easy to appreciate what was involved in these investigations, but some feeling for their complexity is conveyed by the cartoon drawn by David Judd (see credits.)

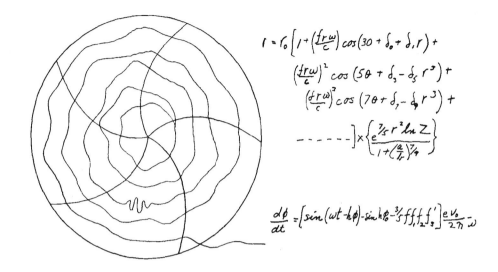

proposals were received: the two cyclotron proposals were rated above the tandem proposals, but MSU's was rated second behind that of the University of Michigan. Then it was learned that the U of M cyclotron had been made a part of the Rivers and Harbors bill, outside of AEC, and that it probably would be approved. It passed and was most likely funded with monies remaining to be expended in FY-1960. This earmark left MSU apparently at the top of the cyclotron queue, but the AEC informed MSU that the results still looked dark for FY-1962 because a tightly constrained budget would probably not permit the construction of any new accelerators. It supported the MSU plan to submit an addendum to the 1958 proposal with an emphasis on uniqueness[51] and suggested that any AEC Division of Research recommendation would be made in July.

The updated proposal to the AEC consisted of the 1958 proposal, an addendum stressing the improved extraction calculations, additional details of experiments, especially for proton-induced direct reactions, and a description of new hires. It was submitted to the AEC in June 1960. Among the new hires, Hugh McManus strengthened the rationale for the direct reaction part of the experimental program. Unfortunately, the new AEC proposal met the same fate as the 1958 proposal. It again received strong technical reviews but no funding, mainly a result of generally tight federal funding. It also appeared that the AEC gave priority to funding more-established programs.

At that point an administrative decision was made to submit proposals to the National Science Foundation (NSF) and to the Office of Naval Research (ONR); the NSF proposal was submitted in December 1960 and the ONR proposal on February 8, 1961. The proposals began with a description of the history of proposals submitted to the AEC, relied on the previous AEC proposals for technical details, and presented a slightly revised budget request. The basic qualitative argument was:

> the proposed cyclotron has been subjected to extremely thorough study [see fig. 13]; roughly four to ten times more is known about the orbits than in any other cyclotron built, in construction, or in design. A design utilizing a novel resonant extraction system has been formulated with all features tailored to maximize beam precision. The studies have recently reached a state of completeness sufficient to allow a quantitative performance estimate for the proposed cyclotron. The figures show a startling improvement as compared with present cyclotrons; the MSU. machine is predicted to extend Van de Graaff precision to a new energy range and with far greater intensity.[52]

In the hope of moving ahead more rapidly, Blosser and Muelder met with Glenn Seaborg, Chair of the AEC on March 16, 1961. He indicated that funding of the 1960 proposal would not be possible,

ACCELERATOR PROPOSALS DURING THE K50 ERA	
AEC	K50 Proposal, December 1958
AEC	K50 Addendum, June 1960
NSF*	K50 Proposal, December 1960
Office of Naval Research	K50 Proposal, February 1961
AEC	Trans-Uranic Facility, May 1969
NSF*	Prototype SC Cyclotron Magnet, July 1974
NSF	Coupled SC Cyclotrons, September 1976
NSF*	K500 Phase I, May 1977
Facilities Subcommittee NSAC	Coupled SC Cyclotrons, February 1978
AEC*	Conceptual Design Report, Phase II—K1200 NSCL, December 1978

*Proposal was successful.

but he did not close the door on funding of $300,000–350,000 to purchase the cyclotron magnet. On March 17, 1961, they received somewhat more encouraging news in a meeting with J. Howard McMillen, Program Director for Physics at NSF.[53] They were told that the proposal was still being seriously considered for support, but no guarantees were possible.

Then NSF laid out two possibilities for MSU consideration: (1) Could MSU proceed on the following grounds: MSU would make a commitment to begin construction of the facility upon receipt of an initial payment of $500,000 from NSF. NSF would make a declaration of intent to pay $500,000 in the two ensuing years or a total of $1.5 million, assuming funds were available. MSU was asked to explore these options. Dr. McMillen added that he did not know whether this could be done legally in his own organization. (2) Apparently reflecting the fact that the review panel was concerned about MSU's lack of experience, MSU was asked whether it would consider undertaking studies for a proposal for a facility in a lower energy range, the thought being that MSU needed experience and time to build up an experimental staff.

MSU responded to these possibilities in a letter from Vice President Muelder[54] on April 19, 1961, stating (the most salient points are quoted here):

- The MSU Board of Trustees unanimously resolved at its March 31 meeting to proceed with construction of the cyclotron building (and other University contributions as indicated in our proposal) if NSF is able to make available an initial allocation of funds adequate to permit reasonable progress on scientific problems, and, in addition, to indicate an intention to carry the project to completion in subsequent years.

- A program of funding which provided approximately $500,000 in each of three successive fiscal years would fit reasonably well with anticipated scientific progress. As contrasted with the funding program outlined in our proposal, the above schedule would extend the construction period by approximately four to six months to roughly 30 months total.

- The *minimum* first year funding adequate to allow reasonable scientific progress is $335,000. This amount of funds would permit procurement of the magnet, coils, power supplies, cooling equipment, and field measuring equipment. Work on field measurements and correcting adjustments could therefore proceed. If first year funds are reduced to this level, it would be desirable to have the second year funds increased correspondingly so that extension of the construction period would be minimized.

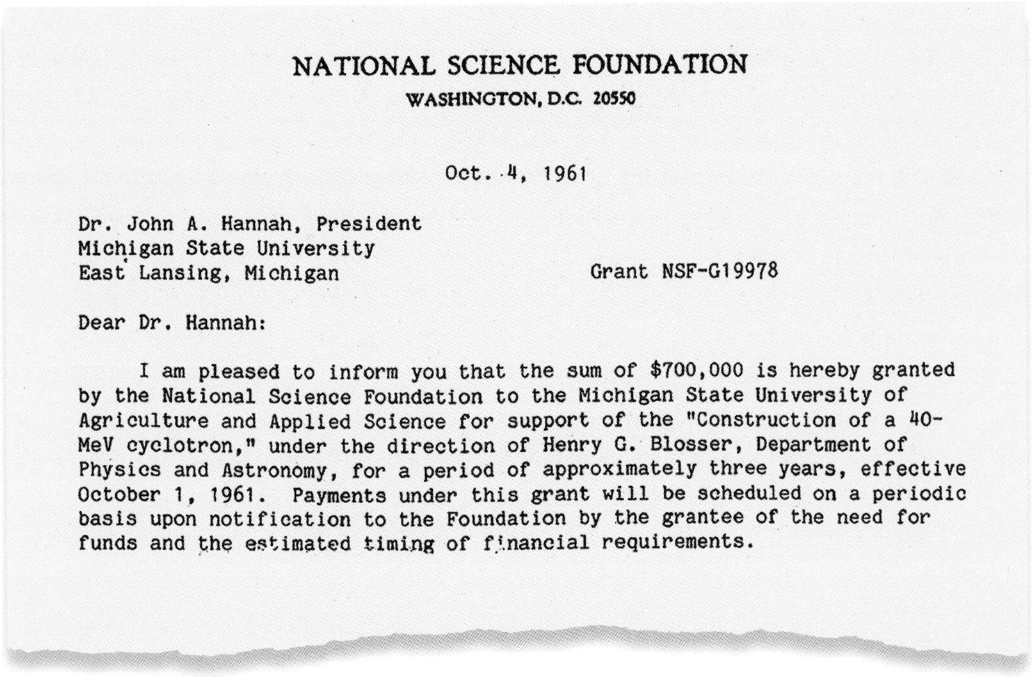

NATIONAL SCIENCE FOUNDATION
WASHINGTON, D.C. 20550

Oct. 4, 1961

Dr. John A. Hannah, President
Michigan State University
East Lansing, Michigan   Grant NSF-G19978

Dear Dr. Hannah:

    I am pleased to inform you that the sum of $700,000 is hereby granted
by the National Science Foundation to the Michigan State University of
Agriculture and Applied Science for support of the "Construction of a 40-
MeV cyclotron," under the direction of Henry G. Blosser, Department of
Physics and Astronomy, for a period of approximately three years, effective
October 1, 1961. Payments under this grant will be scheduled on a periodic
basis upon notification to the Foundation by the grantee of the need for
funds and the estimated timing of financial requirements.

FIGURE 14. The Award Letter from the NSF for construction of the K50. About three years after the submission of the first proposal to AEC and the submission to various federal agencies of four proposals for the facility, it was a great boost for laboratory spirit to finally be able to construct rather than compute and write. It took three years for the funds to arrive and less than three and one half years to complete the project and obtain first beam.

- There appear to be no scientifically reasonable design changes which would lead to substantial reduction in the cost of the cyclotron.

Apparently, AEC and NSF were in close touch concerning MSU's proposals to their institutions. The first indication that NSF had approved the MSU proposal came in a letter dated September 29, 1961, from AEC's Division of Research, stating that MSU's AEC proposal was considered withdrawn, unless AEC heard otherwise from MSU because it had been approved by NSF.[55] At that point, MSU had not heard from NSF and held off withdrawing the AEC proposal until it did. The NSF notice of approval (see fig. 14) was sent from NSF on October 4, 1961,[56] and on November 8, 1961, MSU withdrew its AEC proposal. The NSF grant was for $700,000, with verbal promises of sufficient funds to complete the facility in later years. The amounts actually received were $700,000, $473,000, and $240,000.

FIGURE 15. Milton E. Muelder was dean of the College of Science and Arts (1952–59) and MSU's first vice president for research and dean of the Graduate School (1959–75). He was deeply involved with all aspects of MSU's interactions with AEC and NSF.

The MSU administration was deeply involved in the overall push to obtain funding. Henry Blosser, Dean and Vice President Milton Muelder (see fig. 15), and President Hannah had many contacts with AEC officials, including visits with AEC Secretaries Glenn Seaborg and John McCone, and NSF officials. Most members of the Michigan Congressional delegation were contacted and asked to lobby the AEC. The impression given by these letters is that the campaign was optimistic, almost naïve, considering the countervailing forces. However, it showed great persistence and determination to build the MSU facility and to establish MSU as a player on the national nuclear science scene. And, in the end, it succeeded. Persistence paid! A lesson the laboratory would relearn and apply in the future.

By the end of 1961, the emphasis had shifted from design and raising money to the tasks of recruiting a construction staff and a nuclear science staff and of building the Cyclotron Laboratory and the cyclotron itself. Two new accelerator physicists, Martin Reiser and James Butler, arrived in 1961. With these additions, the accelerator building team was in place. Butler was expected to deal with laboratory administration, while Gordon and Arnette did purely theoretical design calculations. Blosser was also an expert in design calculations but devoted most of his time to the construction aspects of the laboratory. Reiser played an important role in designing the central region of the cyclotron; his design was crucial to providing the precise beams that would be the forte of the new MSU K50 cyclotron. With foresight and optimism, the project team had designed items with long delivery times even before the NSF grant arrived. Construction contracts for the cyclotron building and the orders for the main magnet were submitted in October 1962.[57]

Building the Cyclotron Laboratory, 1963–65

The delay in funding had both negative and positive aspects. The project itself and the science that would be done with it were delayed, but the additional time was spent in fine-tuning the design. The K50 cyclotron, as now conceived, had an expected performance significantly better than that initially proposed. A proposal for operating and equipment support submitted to NSF in July 1963 enumerated these improvements:

- The maximum energy was increased by 20 percent as a result of improved design of the magnet pole tips and revision of the details of the resonant extraction system.
- The frequency range of the radio frequency system was increased and the electrode structure redesigned to allow acceleration on even as well as odd multiples of the radio frequency. This made it possible to accelerate deuterons and alpha particles as well as protons with lower energy (e.g., 10 MeV).
- The cyclotron was repositioned so the median plane was horizontal rather than vertical, improving the optical properties of the extracted beam.

The Cyclotron Laboratory building had originally been proposed as an addition to the

NAMES OF THE CYCLOTRON LABORATORY

- *MSU Cyclotron Laboratory or CL*

  Initial name beginning in 1963 but still used to refer to the institution generically

SUCCEEDING OFFICIAL NAMES

- *Michigan State University/National Science Foundation Heavy-Ion Laboratory*

  October 1, 1977 to January 21, 1980
- *National Superconducting Cyclotron Laboratory (NSCL)*

  January 22, 1980 to February 28, 2015
- *FRIB Laboratory*

  March 1, 2015 to present

Physics-Mathematics Building, now the Psychology Building, on MSU's North Campus. But when the time came for a final siting decision, President Hannah argued that this area was too cramped, so other sites were investigated. According to Henry Blosser these decisions were made at early morning meetings in the Kellogg Center, with a campus map in view, followed by walks to promising sites.

The choice of a site near the intersection of Shaw Lane and Bogue Street (see fig. 16) had both short- and long-term consequences. In the short term it resulted in a one-story instead of two-story laboratory with much more experimental space. For the longer term, the site, extending to Shaw Lane on the north, Bogue Street on the east, and Wilson Road on the south, was preserved for cyclotron use and provided adequate space for several expansions of the laboratory, including now the Facility for Rare Isotope Beams.

The Physics Department was concerned about the separation of the nuclear physics group from the rest of the Physics Department, but acquiesced when it appeared possible that the department could have a new building on South Campus within about five years after the nuclear group's relocation. However, university priorities changed, and that move took more than thirty years.

After receipt of NSF funds late in 1962, construction moved quickly, as is shown in the following snapshots.

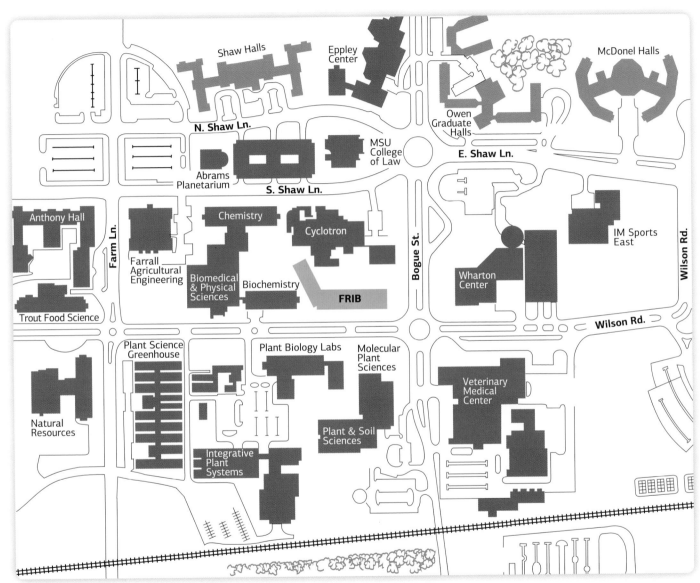

FIGURE 16. Location of the new Cyclotron Laboratory in the modern context, showing FRIB. The growth of the science campus, south of Red Cedar River, has been remarkable. The Cyclotron Laboratory site is bounded by Wilson Road, Shaw Lane, and Bogue Street. The construction of FRIB has led to the closing of Bogue at Shaw Lane.

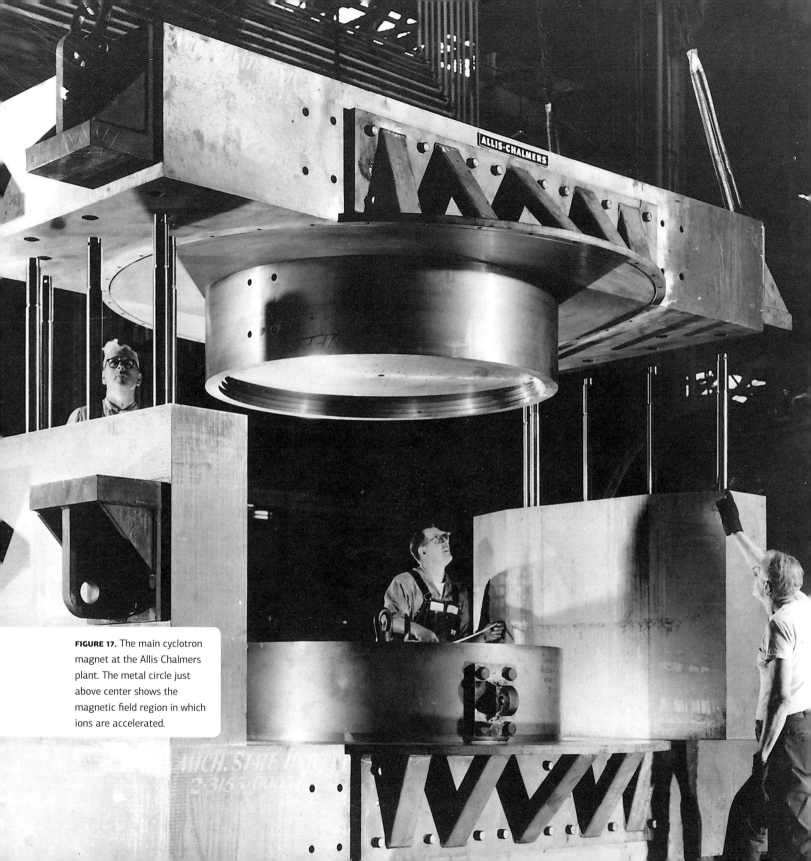

FIGURE 17. The main cyclotron magnet at the Allis Chalmers plant. The metal circle just above center shows the magnetic field region in which ions are accelerated.

■ 1963 Snapshot: The Start-Up Era

Nothing was complete in 1963, but almost everything was underway. Perhaps most important, the new building was completed in October 1963, so equipment could be moved in and assembled as it arrived.

By July 1963, the main magnet was being assembled at the Allis Chalmers plant in Milwaukee (see fig. 17), and the pole tips were scheduled for delivery in August 1963. The main magnet coils were being epoxy coated, and shipment of the completed coils to MSU had begun. The main magnet

FIGURE 18. The Cyclotron Laboratory in 1965. The low section at the front contained offices and laboratories for apparatus preparation. The high bay at the rear contained the cyclotron and the experimental areas.

FIGURE 19. Harold Hilbert assembling the K50 magnet.

power supply was completed and undergoing tests at Pacific Electric Motor in Oakland, California. Late in the summer of 1963, part of the new building (see fig. 18) was released for occupancy, and MSU personnel began assembling the main magnet using a forty-ton bridge crane. The magnet was placed under power early in December 1963, and field measurements at many points, referred to as "field mapping," began.

Such mappings were a recurring part of cyclotron construction at the MSU Cyclotron Laboratory.

Obtaining the proposed high-precision particle beams required that the magnetic fields be known to high precision so the particle orbits could be predicted with sufficient accuracy. This required adding to the main magnet shown in figure 17, a set of eight small coils, "trim coils," to produce the required field.

Another positive development was the arrival of the MSU's Control Data 3600 computer. It was much faster than the MISTIC computer and allowed access from remote terminals. Eventually all the former accelerator programs and new nuclear science programs were adapted to the 3600.

Planning for the experimental areas was in a very preliminary state, consisting basically of lines on a drawing and statements that when the cyclotron was nearing completion, the accelerator group would shift its attention to construction of energy-analyzing magnets and a spectrograph, a magnetic device that sorts particles through different deflections according to their energy, much as a prism sorts light rays by different deflections according to their color. Powerful spectrographs would play an important role at the Cyclotron Laboratory from its beginning to the present.

It had been decided that most of the shielding would consist of stacked concrete blocks, small enough to allow simple future reconfiguration of experimental rooms as needs changed, using undergraduate labor. The building design provided a number of target stations in separately shielded rooms so that a succeeding experiment could be prepared while another was in progress.

There had also been major additions to the research staff. In addition to Arnette, Blosser, Butler, Gordon, and Reiser (who had arrived previously), Assistant Professors Walter Benenson, and William P. Johnson became the first MSU nuclear physicists who had past experience with accelerators. These and later faculty hires, and the time they spent at the laboratory, are summarized in figures 91 and 92. In addition, the laboratory added eleven technical staff members[58] and eight research assistants. A little later Harold Hilbert joined the laboratory as a student worker and eventually played an important role in the laboratory, becoming project engineer in charge of the laboratory building and maintenance of apparatus. One of his earliest tasks is shown in figure 19.

Writing of proposals began to assume major importance for this growing staff. During the early years, proposals for operating funds had to be written every year and funds for equipment had to be proposed separately. Separate proposals to the NSF for operation during 1964, and for shielding, electronics, scattering chambers, a spectrometer, target preparation, and beam handling equipment were submitted and funded.[59]

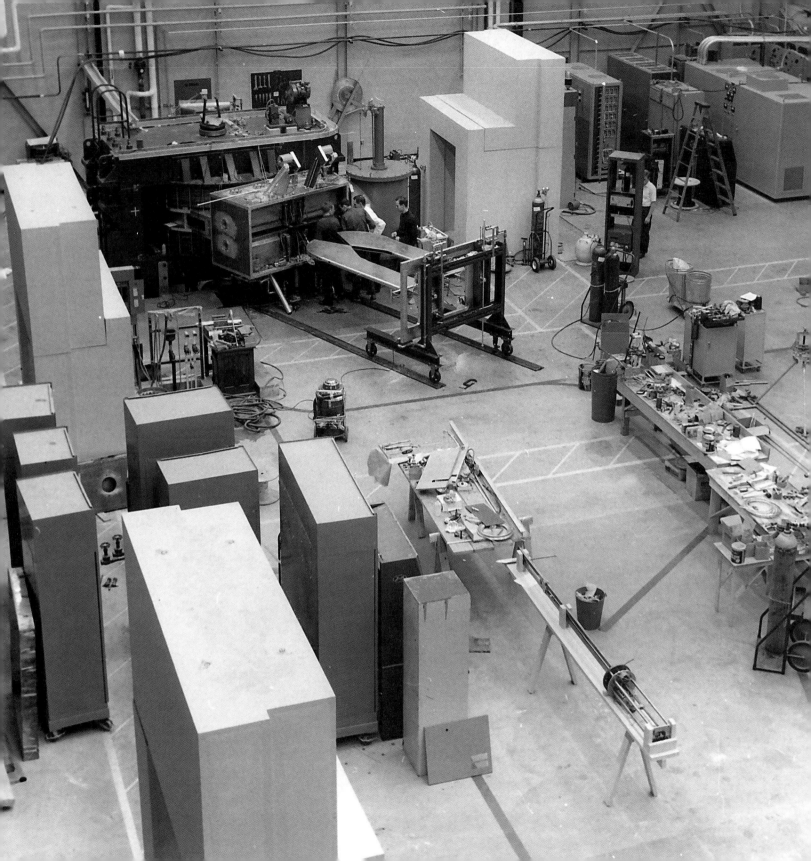

■ 1964 Snapshot: Technical Challenges and Development

A crucial issue was determining the magnetic field. The first mapping had shown it to be of exceptionally high quality. As designed, the magnet produced a field with three high-field regions and three low-field regions going around the magnet reflecting its three sectors. It was shown that field variations at twice this rate, six highs and lows, so-called first-harmonic fields, were very small. These first-harmonic variations would, if large, greatly perturb the particle orbits. Unexpected pluses were that the iron permeability was larger and the coil resistance smaller than had been assumed in the model studies. Together these increased the maximum proton energy to 56 MeV. After the trim coils that perfected the magnetic field were delivered on July 20, 1964, they were installed and their magnetic effects were mapped for each of the eight coils at four different main field excitations.

It was important to control the behavior of the beam in the central region during the first few orbits. Various slits and apertures located in the central few centimeters of the cyclotron were used to minimize the space and time interval for which particles emitted from the ion source at the center of the cyclotron would be accelerated. This in turn determined the beam quality. Learning where to place these apertures was complicated and could, at that time, only be done by experiments in a model of the central regions containing a conducting (liquid) electrolyte. This was a time-consuming process. The design of the resonant system for single-turn extraction earlier studied theoretically by Blosser and Gordon could also be completed now that one knew the actual magnetic fields with precision.

After achieving a leak-free vacuum system in late October 1964, in November and December attention was devoted to the installation of the complex radio frequency system. The 250,000-watt amplifier that provided the radio frequency power was a major component. Large folding copper panels that moved during the changes in frequency needed to obtain different particle energies were also installed, as were the copper-coated dees. The word "dees" is a relic from the early Lawrence cyclotron, where the electrodes were indeed "D" shaped. In a modern cyclotron, the geometry is much more complex, as we see in figure 20, but the name continues.

Although this may sound simple, cyclotron radio frequency systems are not easy to construct and operate. They were a major recurring problem for this and all following generations of cyclotrons at the Cyclotron Laboratory. For the K50, the water connections necessary to cool the movable

FIGURE 20 (*opposite*). K50 cyclotron assembly area. In the upper middle are the dees, far from D-shaped, ready to be inserted into the cyclotron radio frequency cavity. The K50 magnet is behind the dees. The tall white structures hold the doors that will eventually provide entry through the (not-yet-built) shielding walls. The four-foot-thick doors reside beneath the floor and are lifted into place by large hydraulic jacks.

panels and dees were a technical nightmare. They were a source of water leaks that took time to fix and that recurred under operational (heating and moving) stress, plaguing the cyclotron's early operation.

A proposal to NSF[60] requested funds for operation, not for equipment. The lab's hands were full installing the equipment on hand.

■ 1965 Snapshot: Success

In early January 1965, the completed dees were being mounted, and the remainder of the month was spent completing all of the many connections from the cyclotron to the surrounding world: electrical control lines, water lines, remote drives for radio frequency tuning, beam probes, the source of ions to be accelerated, its power and control lines, and many more small items.

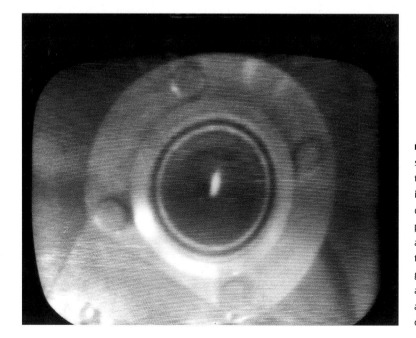

FIGURE 21. Beam out! A simple way of observing the beam is to let it impinge on a piece of quartz located as near as possible to the cyclotron, as was done here. Often this low-tech approach gives the most immediate and useful information about the complex shapes of accelerator beams.

FIGURE 22. Those principally responsible for the construction of the K50 cyclotron. *From left:* Morton M. Gordon (theoretical development), Henry Blosser (magnet construction and overall director), William Johnson (radio frequency system), and Martin Reiser (central region design).

By the evening of January 30, 1965, all was in place, and in front of a large crowd of the interested —students, wives, children, and girlfriends—the startup began. The magnet, trim coils, and ion source were turned on, and the radio frequency warmup began. But in a few minutes a severe internal water leak interrupted the process.

Similar problems occurred for the next ten days, and by a February 11 startup attempt, not one visitor was present. This attempt went exceptionally smoothly. With the computed settings, the beam was soon at the maximum radius (see fig. 21). To quote from a review of the evening:[61]

> While the simple thrill of having the machine operate dominated the thinking of all present, the occasion
> was, in fact, considerably more significant; the machine not only worked but worked with all controls set
> at pre-computed values, thereby establishing that a vast amount of detailed analysis and calculations

had been performed without error and that all fundamental features of the design were sound. The importance of this result is difficult to overemphasize. The ability to reliably calculate beam behavior in an accelerator immediately makes possible a host of development and improvement studies which could not be considered within the framework of previous empirical development techniques due to excessive cost in both money and time. The corollary ability to undertake major accelerator projects with advance assurance on performance is also an important benefit.

Less than three-and-a-half years after funding was received in October 1961, the cyclotron had achieved initial operation. This was a remarkable achievement, especially since it took place in a university with little experience in handling significant technical construction projects and in a department that, at the beginning of the project, had essentially none of the infrastructure necessary to build an accelerator. Figure 22 shows those principally responsible for the construction of the cyclotron.

PERSISTENCE PERSONIFIED

The source of this success must be attributed, principally, to Henry Blosser, who had become a force to be reckoned with. He combined strong technical skills and attention to detail with impatience and an ability to attract and work with talented people, some of whom had strong personality quirks. He was willing to confront anyone—including MSU's President Hannah—whose actions, or lack thereof, he felt stood in the way of the success of the cyclotron project. For example, considering lack of parking at the new laboratory, which he considered an obstacle to morale, Blosser wrote the following to President Hannah:

> I submit that there is something exceedingly inept about a policy which will furnish $250 to furnish a man's office with Steelcase furniture and not spend $100 to give him a parking place from which he can get to that office without walking in deep mud or down the middle of a busy thoroughfare for an extended distance. I can state with certainty that every staff member in this building would overwhelmingly prefer wooden furniture and a parking place than steel furniture and mud. I can also state with certainty that it is excruciatingly irritating to make an inadvertent step in which the mud goes over one's shoe tops.[62]

Hannah sent the Secretary of the Board of Trustees Jack Breslin to talk to Blosser, and he didn't get his parking immediately.

During the preparation of this account I spoke informally with many people about Blosser. Those who were in mid-level campus administration at that time told me several times that the word was "Don't say no to Henry Blosser." In his interactions with others in the MSU and NSF administrative chain, he was demanding or charming depending on his assessment of that person's reactions. His success in bringing a welcome project to MSU had raised his status with President Hannah and, for him, many bureaucratic procedures were expedited.

After this "great success," the plan was to map out the operating space by changing the energy in intervals up to 56 MeV. Indeed, the first step to 33 MeV went smoothly. But then, as is typical in a pioneering project, reality set in as design weaknesses showed up when the machine encountered the stresses of higher energies.

It was first found that the radio frequency amplifier tubes were not adequately shielded from magnetic fields. Then it was discovered that the radio frequency "trimming" electrodes, used to make small changes in the geometry when heating or other effects changed dimensions, had to be redesigned and replaced. This three-month operation caused the largest delay. It was only by midyear that operation was almost routine—an eighty-hours-a-week schedule with about half of the time dedicated to nuclear science research and half to cyclotron improvement studies.

Much work, most importantly installing the so-called conventional (resonant) extraction system, remained to complete the cyclotron.

The early beams were obtained by accelerating negative hydrogen ions made up of a proton plus two electrons. The proton beam was then extracted by passing these ions through a thin carbon foil located at the extraction point. This stripped off the two electrons, changed the ion's curvature in the magnetic field, and directed it into the external beam pipe. It was simple and it worked, but had great disadvantages: one could not obtain truly high-quality beams and one could only accelerate protons because other elements did not form sufficiently stable negative ions. To accelerate positive ions it was necessary to use the resonant extraction system that had been an important part of the K50 proposal to the federal agencies.

The devices required for this, an electrostatic deflector and a magnetic channel, had a simple purpose: they slightly changed the direction of the orbiting beam particles and directed them into the external beam pipe. But both devices presented challenging construction and operating issues. The laboratory had built an electrostatic deflector and had an advanced design for the magnetic channel, but it would be some time before they were in regular use.

Ion source limitations were also important. It was difficult to build reliable and intense sources of heavy ions, so what was to have been a heavy-ion cyclotron accelerated only light ions until the late 1970s.

Building the cyclotron and its initial operation had dominated the attention of the laboratory staff, but as completion of the cyclotron came into view it became clear how much remained to be done to have a Cyclotron Laboratory and not just a cyclotron. One had to build a beam transport system to carry the beam to the experimental targets and detectors, to design and build detection apparatus, and to procure a computer system and related data-analysis equipment. The experimenters, many trained on other types of accelerators, also had to learn to use the new facility.

■ Building a Corps of Experimentalists

With the end of the accelerator-dominated era of cyclotron construction, a nuclear science–dominated era began and most laboratory effort was devoted to improving the detection apparatus and doing experiments with it. All those most deeply involved in cyclotron construction—Henry Blosser, Mort Gordon, Thelma Arnette, and Bill Johnson (Reiser and Butler had moved on)—had assumed that they would return to doing nuclear physics. Although there was movement in this direction, in the end they devoted most of their efforts to accelerator developments. That, in turn, had a major impact on the future of the laboratory because building the non-accelerator aspects of the laboratory relied on the nuclear experimentalists and theorists.

It is a surprise to some that physicists do not all do experiments. Rather there are those, experimentalists, who are experimenters and do experiments aimed at understanding nuclear phenomena. Other experimentalists are mainly interested in developing accelerators or other instrumentation used to perform experiments, as were the initial MSU hires. There are also theorists, who mainly work on paper or with computers, and help to provide the background and methods for interpreting and understanding the nature of nuclei and the results of experiments.

FIGURE 23. Nuclear physics faculty who were present before or shortly after the commissioning of the K50 and who spent much of the remainder of their careers at MSU; they were a dominant influence on the physics performed at the laboratory and on its reputation. *Clockwise, from top left*: Hugh McManus (arrived 1960), Walter Benenson (1963), Aaron Galonsky (1964), Ed Kashy (1964), Sam Austin (1965), Gary Crawley (1965 as a Research Associate, and 1967 as faculty). All but McManus are experimentalists.

Fortunately, owing to an intensive recruiting effort, the nucleus of a group of young experimentalists (see fig. 23) was on site.

- Walter Benenson (PhD 1962, University of Wisconsin) was the first to arrive after a year as a Postdoctoral Fellow at Strasbourg;
- Aaron Galonsky (PhD 1954, University of Wisconsin) had spent five years at ORNL and five years as group leader at MURA;
- Ed Kashy (PhD 1959, Rice University) had spent five years as an NSF Postdoctoral Fellow and an instructor at MIT;
- Sam Austin (PhD 1960, University of Wisconsin) had spent an NSF Postdoctoral year at Oxford University and four years on the faculty at Stanford;
- Gary Crawley (PhD 1965, Princeton University) spent two years as a research associate at MSU before joining the faculty.

In some ways this was an unusual group. They all came from universities with much stronger research reputations, in general and in nuclear science, than MSU, and some were not experienced in using cyclotrons. However, they all had an inclination to participate in the development of experimental equipment, and they fit in well with the build-it-yourself attitude that was growing in the Cyclotron Laboratory. Most of them spent the rest of their careers at MSU and helped propel the program in nuclear science toward its present reputation.

During the initial experiments, negative-ion extraction produced beams that lacked the quality later achieved using resonant single-turn extraction, but they were adequate for the available experimental apparatus. The earliest beams were focused by a quadrupole magnet and transmitted to a single target position located about ten feet from the cyclotron, as shown in figure 24. Experiments were done in a secondhand, thirty-six-inch-diameter scattering chamber obtained from the University of Rochester. By midyear, data had been taken with a 25 MeV proton beam; the results showed that the beam energy resolution, a measure of the spread in energy of the particles in the beam, was about 0.2 percent, within a factor of two of what was predicted.

This experiment revealed a need for apparatus improvements. The rate of data acquisition showed that relying on the CDC 3600 computer in MSU's Computer Laboratory was inadequate.

FIGURE 24 (*opposite*). The initial experimental area for the K50 cyclotron. The cyclotron itself is at the rear of the picture and the beam line and the Rochester scattering chamber are at the front. Richard Dickinson and Aaron Galonsky are standing next to the large square box that houses the folding panels used to vary the radio frequency and thereby the particle energy.

A local online computer was needed. The array of detection apparatus was also inadequate, and there was little progress on two items that could take advantage of the precision beams anticipated with single-turn extraction from the cyclotron: we lacked a time-of-flight system that would use the cyclotron's superb time resolution (0.2 nsec) to advantage, and a particle spectrometer that would benefit from the precise beams to achieve unprecedented energy resolution (1:10,000).

Given the faculty's need to focus its attention on development of the facility, research publications from cyclotron experiments appeared only slowly. There is a one- to two-year delay in publishing physics papers owing to time spent in analyzing data, writing the paper, referee reviewing, and the publication processes. Only one cyclotron-based abstract and no papers appeared in 1965. However, there were several papers describing the MSU cyclotron, accelerator-physics developments that had come to fruition, and experiments successfully done with radioactive sources.

Several proposals were submitted to the NSF in 1965, including one for operating expenses for the next five years.[63]

Cyclotron Laboratory Research

Now there was a cyclotron, but a cyclotron is not the end of the story. It is a tool for performing research in nuclear science, and an expectation and obligation of a research institution is that it provide to society a product that justifies its funding. In the case of the Cyclotron Laboratory, that product is, first and mainly, basic research into the properties of nuclei and the education of young scientists. Second, but importantly, it attempts to foster applications of its research and infrastructure.

If the laboratory is to remain successful, its research must evolve and address the most interesting forefront problems. The quality of its faculty, students, and staff, as well as its accelerators and detection equipment, must also evolve and improve as higher levels of research accomplishment are sought. These have been goals and preoccupations of the laboratory throughout its existence.

■ Observing Nuclear Phenomena

In the following pages some of the research that has been done in the Cyclotron Laboratory will be described. Becoming acquainted with some basic terminology and concepts will enhance the

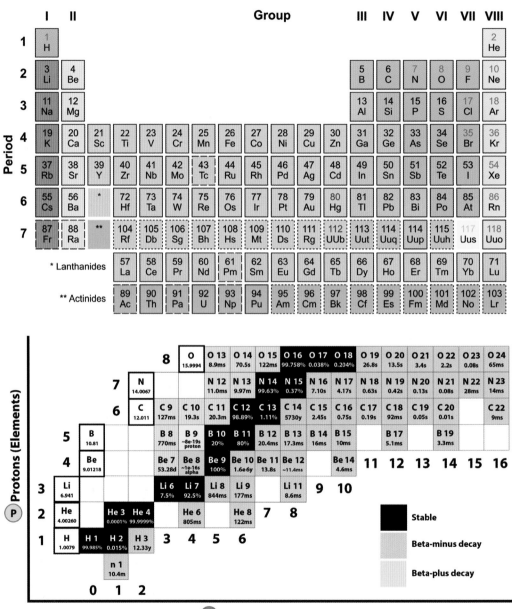

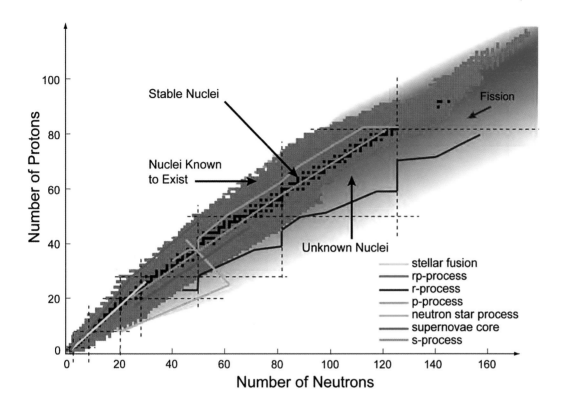

FIGURE 25. Views of elements and isotopes.

Opposite, top: The periodic chart of the elements. Each different element (square) has a different atomic number or number of protons. Elements in a column have similar chemical properties.

Opposite, bottom: Chart of the nuclides (or isotopes). Each square has a given number of protons (Z) and neutrons (N) and a given atomic mass A = Z + N. Squares with the same Z and different N are isotopes of the element Z. In this chart, the element name gives Z and A is specified. Thus, C-9 to C-22 (usually written ^{9}C and ^{22}C) are isotopes of carbon where C-9 has 6 protons and 3 neutrons.

Above: Chart of the nuclides (or isotopes). This is just an expansion of the middle diagram to include all nuclides (isotopes) that, according to present nuclear models, are thought to exist. The black nuclides are stable and are those we see on earth or in stars; the line these trace out is called the "valley of stability." We know many of these nuclides exist, as noted, but there are about 3,000 more that probably exist but have not yet been observed in experiments. The nuclear processes that are likely to make the various nuclides are noted.

understanding of the tools needed for this research and the nature of the research carried out. The following two sections and figure 25 should be considered together.

SPECIAL NUCLEI OFTEN ACCELERATED OR PRODUCTS OF NUCLEAR REACTIONS IN THE K50 ERA

- PROTON (p): One of the two particles (the neutron is the other) from which nuclei are built. It has electric charge and is the nucleus of the hydrogen atom.
- NEUTRON (n): An uncharged particle that is one of two particles (the other is the proton) that comprise nuclei. Neutrons are radioactive, i.e., they decay.
- DEUTERONS (d): Consist of a proton and a neutron and are the nuclei of deuterium or heavy hydrogen.
- TRITONS (t): Consist of two protons and a neutron and are the nuclei of tritium. Tritons are radioactive.
- ALPHA PARTICLES (α): Consist of two neutrons and two protons and are the nuclei of helium atoms.

NUCLEI IN GENERAL

- One must specify the number of protons and neutrons that comprise a particular nucleus (called a nuclide or isotope).
- NOTATION: $^{N+Z}$NAME. Z is the number of protons or atomic number; N is the number of neutrons. The NAME tells us what Z is.
- EXAMPLE: ^{11}C where the name "C" for carbon tells us the atomic number (6 in this case so Z = 6), 11 is the total number of neutrons plus protons. There are then 11 – 6 = 5 neutrons.
- EXAMPLE: ^{2}H is deuterium.

DESCRIBING REACTIONS AMONG ISOTOPES

- The most common notation is: Target (projectile, ejectile) Recoil.
- The target is at rest in the laboratory, the projectile (beam particle) and ejectile are moving rapidly, and the recoil is usually moving slowly. This is the situation for the case where the projectile is much lighter than the target as for K50 research. When doing studies where the

projectile is heavier than the target, as for most studies with rare nuclides, the description is more complex.

- Nuclei are quantum systems and can exist only in discrete energy states. To obtain useful information one must be able to isolate a given state.

- A spectrum shows how the number of particles emitted in a nuclear reaction depends on their energy and therefore on the state that is formed.

- A strong peak (number of observed particles) at a given energy means that the particular quantum state in the nucleus corresponding to that energy is strongly excited.

- EXAMPLE: ^{12}C (p, d) ^{11}C* describes a reaction where an incident proton projectile picks up a neutron from ^{12}C leading to a ^{11}C* recoil and an outgoing deuteron (ejectile). The energy of the outdoing deuteron at a particular angle determines the energy of the quantum state (*) in which ^{11}C is formed.

- Resolution is a measure of the spacing of the states in a nucleus that one can distinguish in a nuclear reaction. For example, if the resolution is 5 keV, one can distinguish reactions leading to two states that are about 5 keV apart. Sometimes one also quotes the resolution as a fraction. If the particle energy is 30,000 keV and the resolution is 3 keV, one can say it is 1 part in 10,000, as in this example. Such resolution is difficult to achieve.

- In figure 27, we show a spectrum where the resolution is 100 keV, typical of the era where we used negative ions, and in figure 31, a spectrum with a resolution of 4.5 keV typical of the positive ion era.

- As nuclei become heavier, the density of levels increases and better resolution is required for their study.

- If one wants to study a nucleus with precision, experimental resolution must be sufficient to distinguish one level from another.

- A projectile may just scatter from the target [elastic scattering: ^{12}C (p, p) ^{12}C]; it may excite it [inelastic scattering: ^{12}C (p, p') ^{12}C*]; it may pick up neutrons or protons from the target or add them to it (transfer reaction); it may add a proton to the target and remove a neutron [charge exchange reaction: ^{12}C (p, n) ^{12}N*]; or it may break up the target [(spallation reaction: ^{12}C (p, pdt)) ^{7}Be]. Each of these processes or their variations provides information about a different aspect of nuclear properties.

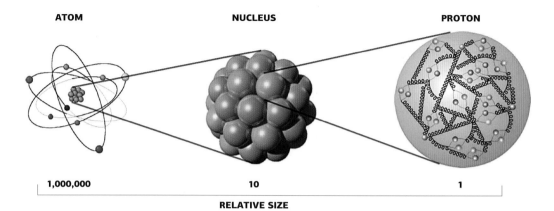

ATOM **NUCLEUS** **PROTON**

1,000,000 **10** **1**

RELATIVE SIZE

FIGURE 26. Relative sizes of atoms, nuclei, and protons.

THE COULOMB BARRIER

Nuclei are positively charged and feel a mutual electrical (Coulomb) repulsion when they are brought near to each other. If that repulsion is strong enough to stop the projectile before it can reach the target, no reaction will occur. This repulsion is called the Coulomb barrier and a reaction will occur only if the projectile has enough energy to overcome it. This is, however, a classical physics statement. In reality, quantum mechanics applies and even when energy is below the barrier, the reaction can still occur but will be improbable.

■ The Proper Tool for the Proper Job

It used to be a truism in mechanical shops that one needed the proper tool for the proper job. In our case, we need to ask, "What is the proper accelerator for the study of nuclei?" This turns out to depend on the size of nuclei.

Figure 26 shows the relative sizes of atoms, nuclei, and protons. Usual pictures of the atom are like that at the left where the nucleus as shown is about 1/5 the size of the atom, but that is only to make the nucleus and the atom visible at the same time. In reality the nucleus is 1/100,000 the size of the atom, though it contains about 99.9 percent of the atom's mass; the atom is mostly very low density space. The proton is about 1/10 the size of the nucleus, but it is made of quarks and gluons, which, as far as we know, are point particles that have vanishingly small dimensions. This is not surprising. The electrons surrounding a nucleus also have vanishingly small dimensions individually, but the electron cloud is relatively large.

Quantum mechanics, through the uncertainty principle, tells us that if we want to study an object of a given size, we need to have a projectile of the correct energy and larger for smaller objects. For nuclei the appropriate energy is in the general range of 1–1,000 MeV. The nature of the projectile is also important, depending on what we want to learn about the nucleus, and whether we want to gently probe the nucleus or to disrupt it.

We conclude: cyclotrons are a proper accelerator for studies of nuclei.

To study the larger atom, you can get by with much smaller energies. To study the behavior of the constituents of protons, the province of particle physicists, you need much higher energies. This means the accelerators need to be much larger; the Large Hadron Collider at CERN in Geneva, Switzerland, is the state-of-the-art device for such studies.

The K50 Era, 1965–79

As the magnitude of the remaining tasks set in, it became clear that it would be two years, at least, before the Cyclotron Lab would leave its metaphorical teen years and become a fully mature facility. But, with the optimism of a youthful faculty and the already excellent capability of the cyclotron beams, a concurrent program of nuclear science experiments and facility development began.

A number of experiments were done with the negative ion, Rochester scattering chamber setup, particularly studying so-called "pickup reactions" in which an ingoing proton picks up a neutron from the target nucleus and the outgoing deuteron is detected. An example, shown in figure 27, is from a publication[64] by Larry Kull, who received the first PhD awarded for work on the K50.[65] Such experiments could only be done on nuclei lighter than calcium. For heavier nuclei the energy levels were often so closely spaced that they could not be separated with the existing 100 keV resolution.

■ Building the Nuclear Science Program, 1965–70

By mid-1965, the layout of experimental systems had been completed and some shielding walls, shown in figures 28 and 29, had been built. The components of the beam lines leading to different

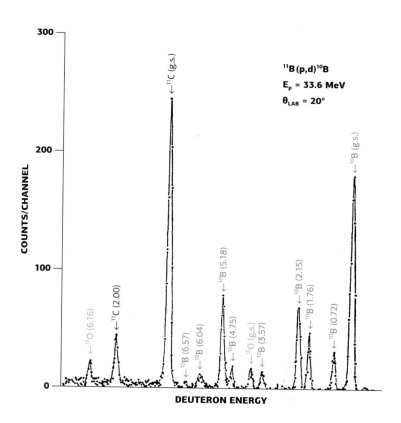

FIGURE 27. A spectrum of deuterons from (p, d) reactions on ^{12}C and ^{11}B, the major constituents of the foil used as a target. The sizes of the peaks corresponding to excitation of different states in ^{11}C and ^{10}B give the relative strengths of individual states. These strengths can be compared to theoretical estimates and can be used to refine these estimates. The resolution is 160 keV FWHM, which means the full width of the peak at one half its maximum value. This is typical of the resolution of these studies.

FIGURE 28 (opposite). An example of shielding walls made of small concrete blocks. This contrasts with the shielding walls in most laboratories which consist of large blocks handled by cranes. But the small block design is much more flexible both in design of the wall and in its reconfiguration. This was done many times, taking advantage of plentiful student labor for the task.

experimental areas—beam pipes, pumps, and diagnostics—had been decided upon. Design of the various bending magnets was complete in some cases and beginning in others, and compact beam-focusing quadrupole magnets[66] were being constructed. Finally, and least advanced, was the required suite of experimental apparatus, including scattering chambers, detectors, spectrographs, computers, and data displays.

The rate of progress was slowed by the demands of nuclear science experiments, the small number of technical personnel, the limited availability of NSF equipment funding, and the ambitious design of the detection equipment. For example, the beam lines were being built with the goal of obtaining exceptional resolution, 1:10,000, in the spectrometer room and at least 1:1,000 elsewhere. Proposals to NSF for a computer system to replace stand-alone digitizers and the CDC 3600 and for additional experimental equipment were funded in March 1966.

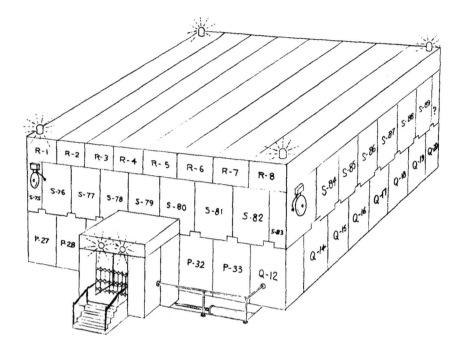

FIGURE 29. "The Cyclotron as Seen by the Health Physicist," a David Judd cartoon. Note the great similarity to the MSU shielding, except for the large blocks and the porch.

A new computer was crucial for coordinating laboratory experimental and theoretical activities. The 1966 computer proposal was for an SDS 930 from Scientific Data Systems, but by the time the grant was received, a more powerful computer, the Sigma 7, was available from that company. It promised to time-share different operations: taking data, analyzing data, and running large programs, all proceeding simultaneously as seen by the users. Although a supervisory program required to time-share was not available, Walter Benenson and a new faculty hire from Bell Labs, Jack Kane, decided to purchase a Sigma 7, serial number 1, and undertake the major effort needed to write this program.

The task was undertaken by two cyclotron laboratory graduate students, John Kopf and P. J. (Phil) Plauger,[67] and the program they wrote, called JANUS,[68] was exceptionally successful. Its time-sharing capability was unique in nuclear science laboratories and rare elsewhere. The Sigma 7 served all laboratory needs[69] for many years, into the early 1980s, although subsidiary computers had to be added to make data-handling more efficient.

Other experimental faculty had joined the laboratory between 1968 and 1972. Most of them, as well as earlier faculty, were involved in apparatus development: *Walter Benenson* for data handling and computers; *Aaron Galonsky* for neutron detectors and a beam swinger; *Charles Gruhn* for building silicon and germanium detectors thick enough to stop our high-energy beams; *Bill Johnson* for K50 operation and improvements; *Ed Kashy* for design and construction of beam lines and development of the high-resolution Enge spectrometer system; *William McHarris* and *Fred Bernthal* for gamma ray detectors; *Jerry Nolen* (see fig. 30) for developing the high-resolution Enge spectrometer system; *R. G. Hamish Robertson* for devising techniques for studying nuclear astrophysics and symmetries in nuclei; and *Sam Austin* for design of a precision scattering chamber. Handling all these tasks with a limited technical staff had positive side effects that led to the development of a laboratory culture that was crucial for the laboratory's further development.

In July 1967, the first beam was transported through the beam analysis system to the new experimental area, and a year later operation with positive ion extraction was routine. It became possible to localize and extract a single orbit (turn) from the cyclotron, a prerequisite for obtaining the promised resolution in energy and time. It also became possible to accelerate deuterons and alpha particles in addition to protons.

The next goal was to develop the Enge spectrometer and beam line system so it could take advantage of the intrinsic precision of the K50 beams. This involved perfecting the technique of dispersion matching that made the position of spectral lines, like those seen in figure 31, independent of the precise beam energy, so a relatively low-resolution beam could produce a high-resolution spectrum.

A resolution meter provided a convenient and accurate way to adjust system parameters to achieve this matching, and feedback mechanisms were used to compensate for drifts in the cyclotron radio frequency voltage and the Enge magnetic field. Given single-turn extraction, these procedures allowed one to image the ion source output aperture on the Enge focal plane and obtain the best resolution[70] that was possible for a given source of ions.

It was unfortunate that one could not build electronic detectors that could match the resolution of the K50-Enge system. Consequently, glass plates covered with photographic emulsions were used to detect the particles in the Enge focal plane when the best resolution was required. These plates then had to be manually scanned in 0.1 mm strips using microscopes. Counting each particle track individually was tiring work, and many undergraduates certainly earned their pay in this effort.

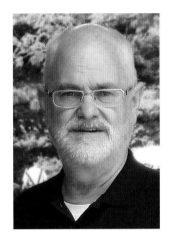

FIGURE 30. Jerry Nolen came to MSU in 1970 following a 1965 PhD at Princeton University, postdoctoral work at Princeton and ANL, and a faculty position at the University of Maryland. He played an important role in development of high-resolution techniques, apparatus, and instrumentation at MSU for the next twenty-two years.

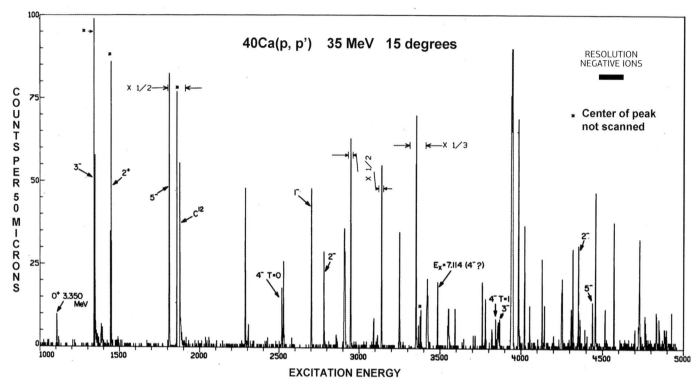

FIGURE 31. Spectrum of protons scattered from a target of ^{40}Ca. Each peak corresponds to the excitation of a single quantum state and its intensity gives information about the nucleus. The resolution in this spectrum is 4.5 keV out of 35,000 keV and is about the best that can be obtained under practical experimental conditions. The heavy bar in the box in the upper right shows the resolution one would have if one used negative ion beams. In most cases one could not cleanly separate the different states and obtain useful information.

By the end of 1970, the Enge split-pole spectrograph, the beam lines leading to it, and the techniques for using it were fully developed and were yielding results with resolutions of 1:10,000 in some experiments, a world-leading figure. Blosser, Crawley, Kashy, and Nolen played the main roles in all these developments, which were then used in many experiments. Figure 31 shows an example, the scattering of protons from ^{40}Ca.[71]

A beam swinger, shown in figure 32, was built somewhat later by Aaron Galonsky, and let one direct the cyclotron beam so as to hit a target from different angles while the particle detector

FIGURE 32. A beam swinger for studying (p, n) charge exchange reactions. The proton beam from the K50 entered from the left in the direction of the red line, was bent by the question mark–shaped swinger magnet, and directed perpendicular to the original beam direction. It hit a target and the outgoing neutrons were detected in a direction perpendicular to the red line and often many meters away. As the magnet rotated about the red line, the angle between the beam and the outgoing neutrons changed, and the corresponding intensity change yielded information about the nucleus being studied.

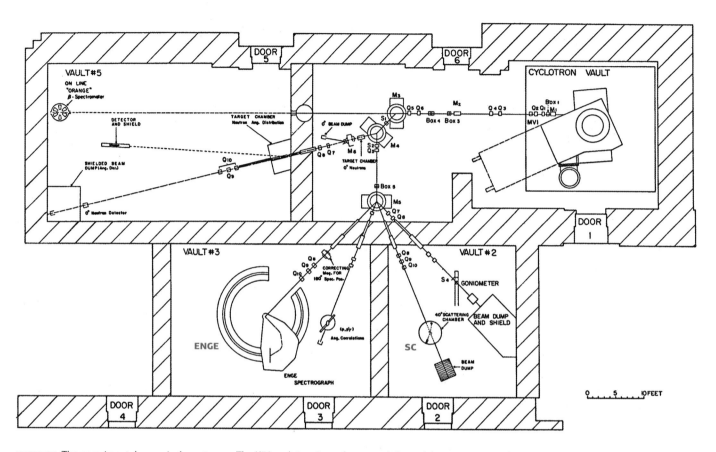

FIGURE 33. The experimental areas in August 1970. The K50 cyclotron is on the upper right and the two most used instruments, the high-resolution Enge spectrograph and the general purpose forty-inch-diameter scattering chamber (SC), are in the lower part of the figure. Other instruments had specific uses and often changed as the experimental program evolved. The shielding walls are the crosshatched areas.

remained stationary. This was particularly useful for studies that involved neutrons because one measured neutron velocity, and thereby energy, by time-of-flight, seeing how long it took to go from the target to the detector. To be accurate, one needed a long flight path, and the detector often had to be far away, many meters, from the target. The swinger made this possible and provided the capability to do high-resolution studies with neutrons that complemented the charged particle experiments with the Enge.

Building space to house this apparatus and the people involved was also developed. Figure 33 shows the three experimental halls housing seven instruments, with the Enge and a precision scattering chamber the most heavily used. A new building addition, completed in 1968, provided additional space for experimental staff and for nearby theory group offices, which greatly facilitated interactions of experimentalists and theorists. By mid-1970, five years after initial operation, the laboratory was approaching its final form as a home for the K50.

■ Gaining Visibility for the Research

A proposal for operating funds to support the K50 program for the next five years was submitted to the NSF in August 1970. When the reviews on the proposal were received, some were surprisingly negative, tending to state that the facility was first rate, but that the science produced was not at the same standard. Having now participated in many such reviews, I believe I understand the basis for these opinions. The visitors saw the facility as it was in late 1970, with all the shiny new apparatus on the floor, displayed by proud builders who emphasized its quality, but it had been available for experiments for only a short time.

When the proposal was written there had been positive-ion extracted beams for less than three years and a high-resolution system for a few months. The reviewers then saw the new equipment, but the less impressive results capable of being achieved with the much less developed apparatus. Nor was the general reputation of the faculty at the level it later achieved. So, while the reviewer's opinions were understandable, they were also a wakeup call: similar opinions five years later would probably have initiated a death spiral for the Cyclotron Laboratory's funding. During that five-year period, many high-quality experiments were completed and published, and this concern proved transitory.

A related issue was the general visibility of Cyclotron Laboratory activities. Although the young faculty members were active and effective, the lab lacked a senior nuclear physicist who could ensure that they were invited to present their work in invited papers at conferences and present seminars and colloquia at other institutions. They were proud even to see someone from another laboratory occasionally present their data at a conference. It takes time to build a reputation.

The laboratory began to work on this problem by pushing colleagues forward for invited talks. Austin, Crawley, and McManus organized the first workshop sponsored by the Cyclotron Laboratory: *The Two-Body Force in Nuclei* held at Gull Lake, Michigan, in September 1971. Most of the important people in nuclear science attended, making it a useful shot in the arm for the laboratory's young facility. The proceedings of the workshop were published by Plenum Press.[72] As time passed, the lab took to heart the need to enhance its visibility and participation in the nuclear science community, and it became a laboratory goal to make sure that the younger researchers were successful.

Of course, the nature of the experiments changed significantly as the experimental capability improved. The early experiments concentrated on light nuclei with mass less than that of oxygen because these studies required resolutions of 100–150 keV that were achievable with negative–ion extraction from the cyclotron and the relatively primitive detection apparatus. With the advent of positive-ion extraction, resolution was three times better and a broader range of nuclei could be studied. Toward the end of this period, use of the new Enge split-pole spectrograph improved resolution by another factor of ten.

This capability was by far the best in the world, comparable to the best obtained with Van de Graaff accelerators, but at higher energies. It opened for study almost all elements and nuclei near the valley of stability. The extensions to new nuclear territory with high resolution and energy yielded qualitatively new insights.

Most experiments were done using proton beams, but later the ability to accelerate particles other than protons opened additional opportunities, as different reactions, sensitive to different nuclear properties, could be used to study these properties efficiently.

The great majority of this work was reported in the best scientific journals in nuclear science: *Physical Review, Physical Review C*, and *Nuclear Physics A*. In 1967, the first article from the laboratory was published in *Physical Review Letters*.[73] The experiment took advantage of the high and variable proton energy (25 to 52 MeV) of the K50 to study how protons and neutrons interact inside nuclei at different energies. This first paper in the most important journal in nuclear science was followed by many other papers as the laboratory established itself as a center for precision research with light ions.

FIGURE 34. Prof. William McHarris operating the K50, a double exposure. For all faculty in this era this was an everyday sight and experience.

FIGURE 35. "The Cyclotron as Seen by the Operator," a David Judd cartoon. Apparently all cyclotron operations are about the same.

The means of operating the cyclotron were unusual for facilities of this scale. There were not enough funds to hire cyclotron operators, so the experimenters ran the cyclotron as well as the beam lines and the detection apparatus. Because there was a substantial learning curve, it was not effective to have temporary people, grad students, and junior postdocs do the machine tuning; that task was most efficiently done by the faculty as shown in figures 34 and 35, while the short-term people concentrated on the apparatus of their own experiments.

As a result the faculty and senior staff became skilled operators and had a strong insight into what worked and what didn't, and into the priorities for improvements. There was some resentment that faculty had to do this "drudgework," but when it was passed on to others, less got done, so in the end it was accepted. Faculty (mainly) operation continued until 1979, when the K50 was shut down, anticipating the operation of a new superconducting cyclotron, the K500. As time passed many procedures became routine, the cyclotron became more reliable, and the overall process became more efficient: an hour was sometimes sufficient to start the cyclotron and bring the beam to the experiment. To be the fastest was a challenge some enjoyed.

We can gain a little insight into the process by examining the information available to the beam tuners, a turn pattern as shown in figure 36. The turns get closer together as one approaches the radius where the beam is extracted. This is a problem, since a single turn must enter the narrow aperture of the electrostatic deflector to extract all of the beam and have precise beam properties. The job of the beam turner was to adjust the magnetic field near the extraction radius so as to induce a resonance that increased the orbit radius at the position of the deflector. It's like pushing a child on a swing at the right frequency so they swing higher and higher. To do this perfectly one had to tune this resonance to increase the turn spacing at the extraction radius

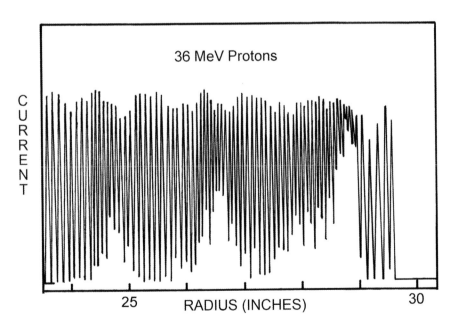

FIGURE 36. The main tool for "tuning" the cyclotron was a graph of orbits in the cyclotron. A thin wire was moved out from the cyclotron center and measured the current at a given radius. Each current peak showed a separate orbit or turn, giving this turn pattern. The orbits in the last six inches before the beam is extracted are shown here. A pattern of this precision was unique to the K50 at the time.

while keeping the spacing of turns uniform nearer the center. There were a lot of knobs to adjust, and some experimenters were better at it than others, but all of us learned to get 100 percent extraction and were proud of it.

■ Administrative Complications

This period also saw the first changes in the leadership of the Cyclotron Laboratory. From September 1966 through August 1967, Henry Blosser was on sabbatical leave at CERN in Geneva, Switzerland, to work on their synchrocyclotron. Aaron Galonsky became acting director of the laboratory and continued as director for three years, after which Blosser reassumed that role.

While he was away, Blosser generated a major controversy with the Dean of the College of Natural Science, R. Byerrum, and the Chair of Physics, S. K. Haynes, over the resolution of a significant cost overrun of grant funds obtained to build the computer system. Owing to a lack of experience with data systems on the required scale and the over-enthusiasm of Jack V. Kane, a data processing visionary, purchasing of all the necessary peripheral equipment for the Sigma 7 was much more expensive than anticipated. That, along with increased overhead rates and electric power rates, generated the overrun.

In addition to cutting other items severely, Blosser's solution to this problem was to eliminate what he called "forced" teaching relief. The laboratory had been paying the Physics Department from NSF grant funds to buy out half of the usual teaching commitment of laboratory faculty members. This practice was common in the department at that time, since it made it cheaper to hire more faculty members, and it was encouraged by the Dean of the College of Natural Science as a way to grow the department faster within the available funds. This practice, at that time, was not forbidden by NSF.

Although Blosser's solution would increase their teaching commitment, faculty members agreed to it for the long-term good of the laboratory. The department and college found it more difficult to accept the loss of funds, and a strong disagreement between the laboratory and both these groups ensued. Although Blosser freely admitted that much of the fault for the overrun was his, he threatened to resign as laboratory director if his solution wasn't accepted, and almost immediately. He got his way.

Somewhat later, in 1971, other perceived difficulties in interactions with the Physics Department led to a shift of the reporting line of the Cyclotron Laboratory from the Physics Department to the College of Natural Science. All faculty and graduate students remained department members, but technical staff and the laboratory director, together with their salaries, were transferred to the Cyclotron Laboratory. So was responsibility for personnel actions, the ability to directly submit proposals, the nuclear physics group's per capita share of secretarial support and seminar budgets, and other small items.

The origin of this decision, made by Provost John Cantlon, lay partly in a desire to simplify handling of laboratory budgets and personnel, but probably most importantly to clear a channel to the Dean for requests for Cyclotron funding and "to make the relationship between the Cyclotron and its users more symmetric,"[74] since a significant part of cyclotron use was from other MSU departments. The flexibility offered by the split was advantageous for the Cyclotron Laboratory, but many in the department resented this action. Bad feelings persisted for many years and, I believe, had negative effects on department support for later Cyclotron Laboratory initiatives.

We did not know it at the time, but Blosser had not yet decided whether he would return to MSU after his sabbatical. Perhaps he did not see at the time how he could, at MSU, duplicate the success of the K50. Whatever the reason, he considered moving to larger laboratories. He applied for the position of Director of SREL, the Space Radiation Effects Laboratory in Newport News, Virginia, but then decided not to accept the position when it was offered.[75] He also applied for a position at the National Accelerator Laboratory (NAL, now Fermilab) in Oak Brook, Illinois. The NAL was in the beginning stages of construction of what was to become the Fermilab accelerator.

Blosser received an offer from NAL and accepted it, planning to begin in January 1968. After his return from sabbatical he spent some time at NAL, but decided that he would not take the position there. He told Wilson that "some very interesting new ideas for a major reworking and improvement of the cyclotron have come up" and that "for one reason or another I just haven't gotten very excited about anything at NAL as yet."[76] Other comments in his letter indicate he felt that it was too late to participate in the main decisions on the structure of the accelerator and that he would be a relatively small part of a very large project guided by a dominant leader, Robert Wilson. Later in his career at MSU, it became clear that Blosser was only truly comfortable in a dominant leadership role.

It is not known what interesting projects at MSU, if any, Blosser actually had in mind when he decided not to go to NAL. At MSU, he became involved in developing the high-resolution capabilities of the Enge spectrograph and participated in his only nuclear science experiment[77] at MSU. He was apparently motivated primarily by the technical challenge of achieving high resolution with the K50, rather than by the nuclear science questions it might answer. His future interests lay in building more powerful accelerators, and by 1969 preliminary designs for a "Trans-Uranic Research Facility" for MSU had been developed.[78]

■ Thinking Bigger—A Transuranic Facility

The MSU Transuranic accelerator was to be capable of accelerating all elements to energies above the Coulomb barrier so that nuclear reactions could take place. Transuranic elements have an atomic number greater than ninety-two, that of uranium. (Among these are recently named elements with atomic numbers 114: flerovium, and 116: livermorium.) One specific goal was to search for the predicted islands of nuclear stability well beyond the range of then known nuclei, and more generally, to determine whether what has been learned from the normal nuclei seen around us could be applied in a new region of rare isotopes. If not, it is a discovery of new phenomena.

The proposed facility consisted of two accelerators: a large cyclotron would further accelerate beams of heavy ions produced in a small injector accelerator. Options considered for the injector were an electrostatic tandem accelerator, a smaller heavy-ion cyclotron, and the existing K50 cyclotron.

A document outlining the nature of the proposed facility was widely circulated in June 1969, but no budget details were included. In mid-July the complete proposal was submitted to NSF and AEC.[79] It requested a total of $18,280,000 (approximately $116,000,000 in 2013 dollars). A later letter from the MSU's Provost and President committed to "a long-term $1 per year lease arrangement of the land on which the facility is to be constructed, providing the agency full rights to operate the facility in accord with your view of the needs of the nation."[80] The K50 cyclotron and its building were included in this offer.

The proposal was held by the AEC for some time, perhaps indicating an intrinsic interest, but was declined[81] on April 2, 1970, on the grounds that the President's FY-1971 budget did not have the funds for such a facility. The AEC stated that the Heavy Ion Linear Accelerator (HILAC) established at the Lawrence Radiation Laboratory (now LBNL) would "likely be the only major high energy, heavy ion accelerator in the United States for the near future . . ." As we shall see, by 1978 they had changed their minds. Although this proposal failed, it was the first indication of the interest of Blosser, the Cyclotron Laboratory, and MSU in establishing MSU as a player on the international scene in nuclear science.

Some negative reviews of the proposal led Blosser to conclude that others interested in such a facility found it advantageous to be strongly critical, and that their negative views, although often uninformed, were given undeserved credibility because they came from national laboratories. He strongly expressed these views to the AEC.[82] His conclusion was that any future proposal for a large accelerator should separate the construction and final site choices. Other laboratories would then be less likely to oppose such a construction-only proposal, hoping they might later receive the completed device without having to build it.

The K50: Its Golden Years, 1970–79

By 1970, reliable operation of the K50 and the Enge spectrograph had become routine, and high-resolution studies of many types of reactions were the trademark of the laboratory. Much of this work was enhanced by collaborations between nuclear experimentalists and theorists. B. Hobson Wildenthal, an experimentalist, in collaboration with theorist B. A. (Alex) Brown developed state-of-the-art shell model descriptions of nuclei and made predictions of nuclear properties for comparison with experiments. Theorists Hugh McManus and George Bertsch developed theoretical infrastructure for studying nuclear structure and reactions, and served as gurus, helping experimentalists plan and decipher the results of experiments.

The shell model of nuclei considers the motion of individual protons or neutrons (collectively, nucleons) in an attractive field. An important object of experiments is to determine how many of the nucleons are in each of the many possible orbits and how they form the quantum levels of the nucleus. The MSU experimental results, when compared with the theoretical shell model calculations, were crucial in elucidating the shell structure of nuclei and the nature of the non-spherical shapes seen in many nuclei. The high-resolution, proton- and neutron-based programs took advantage of the high-quality optics and time definition of the K50 beams. These properties were unequaled anywhere in the world.

FIGURE 37. George Bertsch, a nuclear theorist, received his PhD from Princeton University in 1965, and after appointments at Princeton and MIT, came to MSU in 1971. He later received a Hannah Professorship. He was influential in guiding both the nuclear structure studies and later studies with heavy ions.

■ Expanding the Research Scope

Faculty also developed interests in related fields. Nuclear reactions drive the evolution of stars and produce the elements we see around us, most importantly those required for life—carbon, nitrogen, and oxygen—but almost all elements as well. The signatures of these reactions are often obvious in stellar spectra and provide clues to the processes that underlie the evolution and structure of stars. In this sense, nuclei are:

> The DNA of the Cosmos. Imprinted upon nuclei—by virtue of the special characteristics of the strong, weak and electric forces and the principles of quantum mechanics—are the blueprints for the universe, for the stars and the elements therein, and ultimately for life.[83]

Studies of astrophysical phenomena became another trademark of the Cyclotron Laboratory and one of the drivers for later laboratory accelerator projects, especially the Coupled Cyclotron System (CCS) and FRIB. Graduate student, Helmut Laumer, working with Cary Davids, a postdoc from Caltech, had the first publication[84] and PhD in this field. Later, Laumer returned to the Cyclotron Laboratory and was in charge of the cryogenic systems for the superconducting cyclotrons that were on the horizon.

Other pioneering experiments were carried out using radioactive ^{13}N as a biological tracer. Galonsky and Austin had developed techniques[85] for making uniquely intense sources of radioactive ^{13}N and biologists, Peter Wolk and James Tiedje from MSU's AEC Plant Research Laboratory and Department of Agriculture, had developed techniques for using it in experiments.

PERFORMANCE OF THE K50

- Beams uniquely precise.
- First cyclotron to: (1) achieve single turn extraction and (2) achieve 100 percent extraction efficiency.
- Use phase selection to achieve sharp time selection (200 picosecond timing).
- Use feedback signals from the extracted beam to stabilize radio frequency voltage.
- Directly image the cyclotron's ion source on the focal plane of a high resolution spectrograph, and thereby achieve unprecedented energy resolution (1.5 keV for 30,000 keV protons).

EXPERIMENTS WITH ALGAE AND DE-NITRIFICATION

The experiments with algae were high-speed experiments. ^{13}N has a half-life of ten minutes, so after it was bombarded, one had to extract it from the target and convert it to gaseous nitrogen, run to the Plant Biology Laboratory across Wilson Road, let algae eat it, and then determine into which amino acids it had been incorporated. And all before the ^{13}N had decayed away. After an experiment, we were exhausted.

The first studies examined the sites and metabolic pathways of nitrogen fixation in blue-green algae: radioactive ^{13}N was incorporated in nitrogen gas, fed to algae, and traced though their metabolic processes.[86] Later studies addressed the de-nitrification process that releases nitrogen from nitrate fertilizers into the air as gaseous nitrogen. These studies continued with the K500 cyclotron; thirty-three papers were published by MSU researchers between 1974 and 1981 and served to define the field.[87]

■ Creating a Culture

These were years that set the Cyclotron Laboratory on its future path. The success of the K50 and its associated experimental apparatus gave the laboratory credibility for later expansion. Equally important for the long-term viability of the laboratory was an evolving and, it would turn out, enduring laboratory culture. Some of its aspects arose from Henry Blosser's views but others seem to have arisen from the nature and opinions of the early senior hires and limited early funding that made it difficult to hire accelerator operators and technicians. This resulted in the creation of a laboratory personality built on pride in the lab and a view that the whole is greater than the sum of its parts.

Some of its practices, never formalized, were to: give senior staff strong informal influence on lab decisions; hire high-quality research staff and give them free rein in the choice of research; build forefront experimental apparatus; involve senior staff directly in machine operation and apparatus development; and encourage interactions of experimental and theoretical faculty members.

Many of the faculty interactions played out in the hallways, as all offices were 121 square feet in area, too small to have gatherings, or in Friday meetings at Blosser's house, where everyone sat on the floor, with beers in hand, and did laboratory business. Blosser had a strong personality and often got what he wanted. But often he did not: important items had informal votes that decided issues. Even those who were opposed, including Blosser, bought into these decisions because everyone knew what had been considered and decided.

The shortage of funding had slowed the development of the laboratory and made it difficult to hire specialists, with the result that the structure of the Cyclotron Laboratory was extremely flat, with minimal bureaucracy. There were no experimental groups that had a defined interest and composition. Rather there were ad hoc collaborations formed to attack specific problems. To a large extent, everybody interacted with everybody. This included theorists who were interested in talking to experimenters and helping to guide and analyze experiments.

All in all, this made a stimulating and supportive environment, especially attractive to students. One of them, Robert Doering, a PhD student of Aaron Galonsky, expressed these feelings in an American Institute of Physics Oral History transcript.[88]

> Most of the professors worked pretty much individually, sometimes in pairs, and they might each have one or two students. . . . So, it was very common for them to invite students of other professors to help with their experiments, which was encouraged both ways. Thus, you got to meet everybody pretty quickly and learn a lot of different experimental techniques and co-author papers far outside of your thesis area. That was a great experience. In fact, I wouldn't have minded if time could have magically extended so that I could have kept working in that lab forever. To me, it was an idyllic world, and it was as much fun as anything I've done in my life.

During the early 1970s, the research program at the K50 cyclotron was at its most productive and had a bright future, at least for another decade. But every accelerator has a finite life. Although the K50 was the best of its type in 1965, two labs[89] had built close copies, with MSU help, and other types of accelerators were becoming more competitive. Not too far in the future the K50 would have skimmed the cream of the discoveries that might be made with such a machine, and new experiments would become more difficult, more time consuming and, on average, less informative.

This situation could be postponed by developing unique new experimental apparatus or new accelerator capability, and during this decade the Cyclotron Laboratory did so.[90] But to continue to work at the cutting edge, the laboratory would need a unique new capability. And its development had to begin a decade or so in advance, while the present facility was producing state-of-the-art results. It takes that long, at least, to move from concept, to funding, to apparatus on the floor, to reliable operation.

Fortunately, an opportunity soon presented itself and led eventually to the construction of a K500 superconducting cyclotron and closing down the operation of the K50 in July 1979.

Beginning of the Superconducting Era

There was, in the early 1970s, an enthusiasm among nuclear physicists for the science that could be learned by studying collisions with heavy-ion beams. One could use these collisions to heat and compress nuclear material and study its properties in detail. There was also federal activity that encouraged such research. Blosser had served on an "Ad Hoc Panel on Heavy Ion Facilities" appointed by the National Academy of Sciences-National Research Council at the request of the AEC to study the importance of heavy-ion facilities in nuclear science.[91]

■ Opportunity in a New Field

In their 1974 report, the Panel recommended that a system consisting of a large electrostatic accelerator injecting an isochronous cyclotron could "take full advantage of the opportunities in heavy-ion science." The advantage a two-stage facility offers is that the energy a cyclotron produces increases with the square of the charge of the accelerated ion. The ion from the injector will be stripped of many electrons during its injection into the second-stage cyclotron and have a high charge, making high energies possible. The Panel considered use of superconducting cyclotrons as the second stage, but felt that it was too early to judge their likely performance and cost. There was, however, strong support for pursuing the necessary research and development.

This presented an opportunity for an accelerator builder like Henry Blosser, who enjoyed the challenge of building a bigger and better accelerator but had to justify such a project in terms of the science it would produce.

Building a cyclotron with a superconducting magnet coil, instead of a room-temperature coil, appeared to be an obvious development direction and one that built on the MSU Cyclotron Laboratory's core expertise. A superconductor loses all resistance to the flow of electricity when it is cooled to a temperature near absolute zero, the lowest possible temperature; this eliminates the energy normally consumed when current flows through a room-temperature magnet coil. The major decrease in power consumption is one advantage, and in addition, the higher fields that can be achieved allow one to make a much smaller and cheaper cyclotron for a given energy. The rest of the cyclotron would remain at room temperature.

But there were opposing arguments. Blosser thought that the small size of a superconducting machine would cause difficulties with beam quality and extraction. In a comment he made at the 1972 Cyclotron Conference in Vancouver,[92] while reviewing prospects for future cyclotrons, he concluded: "Superconductivity then seems unlikely to make a contribution to cyclotrons in the foreseeable future primarily because there is no overriding problem which would thereby be solved such as is the case for synchrotrons and linacs."

His opinion soon changed, however, and he described the evolution of his thinking during an address at the dedication of the National Superconducting Cyclotron Laboratory in 1982.[93]

These thoughts had in fact first occurred to us as far back as 1963—an early graduate student in the laboratory, Dr. Richard Berg studied the design of such a device in a thesis project. This particular experience was however quite discouraging. The quality of the superconducting materials which we were able to obtain was very poor, the materials were unpredictable and unreliable, and so the idea was put aside. In February 1973, another very important event occurred, namely Drs. Bigham, Fraser, and Snyder from the Atomic Energy of Canada Laboratory in Chalk River, Ontario, came to visit. . . . Bigham and his collaborators had become interested in the prospect of building a superconducting cyclotron and had quickly come to the conclusion that reliable, predictable conductors had by that time become readily available. In the fall of 1973, this group issued a very significant technical report[94] which attracted wide attention in the accelerator community. Somehow, though, I was personally not yet ready to see the message, at least neither the February visit nor the receipt of this publication gave me a sense that this was something I should be actively working on myself. That important step occurred later in the fall of 1973, when I spent a very pleasant two weeks visiting the Princeton University Cyclotron Laboratory at the invitation of Professor Rubby Sherr. One afternoon, in the middle of this two weeks, Professor Milton White walked into my office and said, "Wouldn't it be smart to think of increasing the energy of the Princeton cyclotron by replacing the conventional coil with a superconducting coil?" I said something to the effect, "Well, I haven't thought much about that. Let me scratch my head a bit."

■ A Prototype Superconducting Magnet

These events led to the initiation in November 1973[95] of studies for a full-scale prototype superconducting magnet to be built at MSU. Properties of novel magnets are usually studied in small-scale models, but this cannot be done with superconducting systems: scaling to significantly smaller size requires current densities that are not achievable. A partial compensation for the cost of a full-scale prototype is that, if successful, it can then be used as the basis of a cyclotron.

The details of the design were given in a proposal submitted to NSF and AEC in July 1974.[96] The magnet size was chosen to give a bending power of K = 400, adequate for a wide range of heavy-ion science. The superconducting coils were modeled on bubble chamber coils that had been built at Argonne National Laboratory by John Purcell. Purcell and ANL had agreed to "be in charge of the fabrication and testing of the complete coil, cryostat, and refrigerator system."[97] The proposed cost of the magnet was $840,700 of which $450,000 was for the superconducting coil and its support infrastructure. To convert the magnet to a complete cyclotron was stated to require an additional $828,000. This proposal stated: "We specifically stipulate that the magnet will be available for disposition *in any fashion* and *at any time* designated by the funding agency . . ." Arguments for and against siting a major heavy-ion facility at MSU were presented. A possible outcome of granting the proposal was:

> The procedural structure which we then suggest as most effective, namely to build at least the magnet and perhaps the complete cyclotron at Michigan State and then a probable transfer to some other

SUPERCONDUCTING CYCLOTRON MAGNET – K = 500 MeV, K_F = 160 MeV

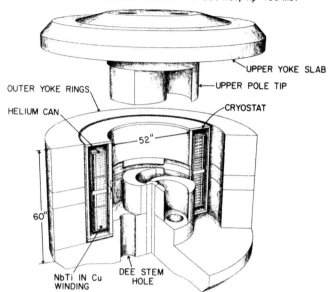

FIGURE 38. Views of the K500 prototype superconducting magnet during its construction.

Top left: A conceptual view. The ring-shaped superconducting coils, divided into four sub coils, are shown. The fine strands of superconducting NbTi alloy are imbedded in a much larger mass of copper. A helium can holds the liquid helium bath that cools the coil, is surrounded by insulation, and is placed in a vacuum cryostat. The entire cryostat is placed in an iron yoke, shaped roughly like a pillbox, which supports the curved magnetic poles and also serves to confine the magnetic field.

Top right: The (almost) assembled magnet. Although the structure is complicated it does not include the many external elements needed to turn a magnet into a cyclotron.

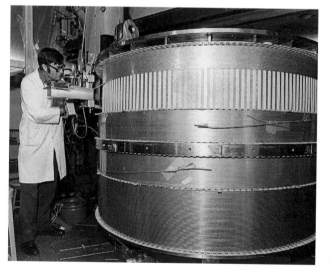

Bottom left: The superconducting coil with Norm Bird. The four sections are visible; the top section is incomplete. In the upper right diagram, it has been inserted into the magnet steel.

laboratory, would be a novel construction format for research cyclotrons in the U.S. with advantages and disadvantages relative to the traditional build-your-own-cyclotron approach.

Blosser thought that the advantages had mainly to do with increasing the probability of funding. He also felt that possession was nine-tenths of the law, and that if built at MSU, it would likely stay there.

The proposal was funded by the NSF at $202,400 in June 1975; additional funds of $800,300 were received in later years. Construction then began (see fig. 38).

Construction of the superconducting coil for the K500 prototype magnet was originally planned as a collaborative project with John Purcell's group at the Argonne National Laboratory, with the coil to be constructed at ANL. However, conflicts with other ANL programs[98] made this impossible, and coil construction was moved to MSU.

In the beginning, MSU's external suppliers had difficulties meeting specifications: the coil bobbin was strongly deformed by welding errors that took time to repair, and the superconducting wire had many flaws, requiring about 1,000 splices and repairs during coil winding. Winding began in October 1976 and was completed in January 1977, only slightly late, and under budget. This made it possible to purchase enough superconducting wire to rewind the coil in case the fixes to the flawed wire proved inadequate.

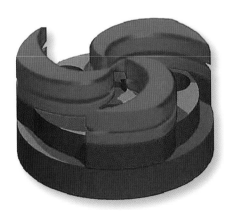
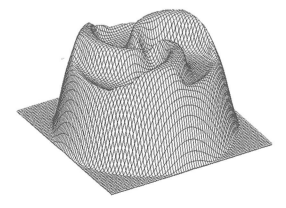

FIGURE 39. *Left:* Schematic view of the iron pole tips of the K500 cyclotron (in red, resting on the supporting iron structure of the magnet). *Right:* A three-dimensional view of the measured magnetic field is shown. The vertical height gives the strength of the field. The effects of the three sectors are evident.

FIGURE 40. The magnet power supply door showing damage caused by an electrical arc, which lasted many minutes. This was a "Fire Department" event.

The magnet was tested to full current on May 26, 1977, and it was found that there was agreement with the predicted magnetic field to within 0.2 percent. This validated the assumption that all the complex iron shapes were fully magnetized so their effects could be well approximated by surface currents. It also greatly simplified the field calculations.[99] A representation of the resulting field is shown in figure 39.

The magnetic fields of a magnet of the size of the prototype store a large amount of energy. To turn the magnet off, one must find a way to remove this energy. This was done by constructing a resistive metal strip that one could place across the leads of the magnet. This resistor is placed in a water bath to prevent it from melting when the energy of the magnet is released.

Normally this takes a long time, but there was a subtle flaw in the design, and during a shutdown, the resistor failed at full field. This produced an open circuit. The resulting high voltage arcs damaged the power supply as shown in figure 40, in a fast dump of all the coil energy. Later, in 1977, there were two events when the coil support links broke during tests. However, the coil survived these inadvertent tests of its robustness with no observable change of its characteristics.

All of these activities made clear that it was important to develop expertise in this new field of heavy-ion science. A first step was to accelerate heavy ions using the K50 cyclotron. Two steps

FIGURE 41. Konrad Gelbke came to the MSU Department of Physics and the Cyclotron Laboratory as Associate Professor in 1977. He has remained at MSU since that time and was promoted to University Distinguished Professor in 1990. He quickly established a prolific and influential research program studying collisions of heavy ions and later served as Director of the laboratory.

FIGURE 42. David K. Scott came to MSU as a Hannah Professor, MSU's highest academic honor, in 1979. In addition to his own influential research, Scott was an eloquent community spokesman, through his personal interactions, review articles, and talks at major conferences, for the value of heavy-ion research in general and the MSU research in particular.

were required: (1) developing the higher, $n = 3$, harmonic acceleration modes for the K50 that are necessary for heavy ions ($n = 3$ means the radio frequency is three times the orbital frequency of the ion), and (2) providing workable ion sources for heavy ions that would work in both the K50 and K500 environments. This work was carried out by Merritt Mallory[100] and his collaborators.

On September 16, 1976, 70 MeV beams of ^{12}C and beams of ^{6}Li were extracted from the K50 cyclotron; a small experimental program with these ions began on September 20. By a year later, a much more extensive list of beams was available, with energies close to those originally proposed to AEC and NSF.

At last, the K50 had become the heavy-ion cyclotron that was conceived in 1958. But in the interim, its unmatched capability as a light-ion facility had helped to redefine what a modern cyclotron could do to advance our understanding of nuclei.

A shift to research with heavy ions followed, both at the K50 with first use of ^{6}Li beams, but mostly at other labs, particularly at the Lawrence Berkeley National Laboratory (LBNL) where more powerful heavy-ion accelerators were available. This new emphasis led to the hiring of physicists with heavy-ion experience: Associate Professor Konrad Gelbke from Heidelberg University via LBNL in 1977, and Hannah Professor David K. Scott from LBNL in 1979.

The Midwestern Collaboration

Whether a large facility would be built at MSU was unclear. The prototype K400 (now K500) superconducting magnet was funded with the explicit understanding that a cyclotron based on that magnet could be moved elsewhere. There was, moreover, a strong candidate for an alternative site at the University of Rochester. It had an existing injector, a large tandem, and MSU had none. With this in mind Blosser was collaborating with Rochester, hoping that they would move their device to MSU. As might be expected, Rochester had the opposite hope.

The possibility that a K500 cyclotron-based facility might be built at MSU was greatly enhanced by the desire of nearby nuclear physicists to have a large accelerator in the Midwest. An early meeting for this purpose had been held on January 11, 1968. Participants from ANL and the AUA (Argonne Universities Association), including three from MSU, met at O'Hare airport in Chicago to delineate desirable specifications of a proposed ANL-AUA Midwest accelerator facility. It would be funded by the AEC and, presumably, located at ANL. As outlined in a letter[101] from W. C. Parkinson of the University of Michigan, the proposed facility would include a state-of-the-art tandem Van de Graaff which would inject a ring cyclotron of unspecified size, presumably around K = 400. To my knowledge, this effort had no positive result.

■ Leveraging the Prototype Magnet

Then, on October 4, 1974, having received information that funding of the prototype superconducting magnet was likely, Henry Blosser proposed[102] a short meeting at East Lansing on October 18, 1974, to "review possibilities for proposing a collaborative facility for the region." As an example, he described an $8,000,000 facility including a K = 400 superconducting cyclotron based on the proposed prototype magnet ($1.5M), a tandem accelerator injector ($3M), experimental equipment ($1.5M), and a building ($2.0M). Eighteen people from eight institutions attended the meeting.

On October 23, 1974, Blosser reported[103] to MSU Provost John Cantlon, that the group (plus three other institutions via letters) had made a tentative decision to proceed with such a proposal but wanted first to determine whether a two-cyclotron system might be superior to a tandem-plus-cyclotron system. He requested conceptual approval to proceed, with the goal of submitting a proposal by January 1975. He also asked for a statement of MSU monetary support (10 percent of construction costs as a possibility) and MSU willingness to designate a facility director as recommended by the scientific sponsors and approved by the funding agency. A large group of sponsors, plus these expressions of support, he felt, should be sufficient to overcome the cost advantage of proposers (Rochester, for example) who possessed an injector.

Soon thereafter, on November 25, 1974, Kirk W. McVoy of the University of Wisconsin circulated to a group of Midwestern nuclear physicists, a letter[104] with the heading "Observations on a Possible

FIGURE 43. A logo showing the concept of the coupled superconducting cyclotrons. Beam is initially accelerated in the K500 cyclotron, is extracted, transported to the K1200, passed through a thin foil that increases its charge, and then accelerated in the K1200 cyclotron.

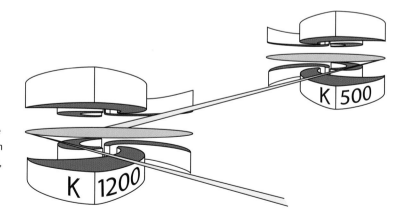

NSF Interest in a New Electrostatic Accelerator for Heavy-Ion Research." He reported impressions, apparently resulting from conversations with NSF officials, that at a cost level of $15–20 million, NSF could provide the construction funds for a joint effort of several universities, provided NSF was convinced of the scientific merit of the science to be done and could find a way to cover operating costs of the new facility. This presumably meant that participating groups would direct their activities to the new facility, saving the operating costs of their own accelerators. Another impression was that Congress and the funding agencies were more interested in funding new facilities than continuing the support of old ones.

A December 17, 1974, letter from Paul Quin,[105] also of Wisconsin, stated his "strong support" for the MSU proposal rather than a large tandem because it provided unique capabilities for heavy-ion research. He also suggested several candidates as outside members of a Board of Sponsors, including W. C. Parkinson of the University of Michigan. It was clear from their letters that the Wisconsin group, especially McVoy and Quin, hoped to play a major role in any proposal but did not have the desire or manpower to lead it.

Given this desire and the interest already expressed, on February 21, 1975, McVoy and Quin circulated a letter[106] suggesting a meeting at O'Hare Airport for further discussions of a regional facility. The positive response to their letter was followed by a one-day meeting on March 8, 1975, with about thirty attendees, chaired by Parkinson. The morning session comprised three presentations: (1) P. Quin on the performance of large tandems, (2) John P. Schiffer of ANL on science accessible to large tandems, and (3) H. G. Blosser on the proposal for a cyclotron-based, heavy-ion system. The afternoon session involved discussions of the siting and organization of a consortium.

A second meeting held on March 28 and 29, 1975, at the University of Wisconsin reviewed the (mostly) heavy-ion science that could be addressed by a higher energy nuclear accelerator. Finally, a third meeting at MSU on April 11 and 12, 1975, was devoted to a review of the characteristics of likely accelerators. There were presentations about possible heavy-ion science at the Indiana University Cyclotron Facility (IUCF); on performance of the upgraded MP tandem at Chalk River Nuclear Laboratories; and on 20–25 million volt tandem Van de Graaff accelerators.

However, the two serious contenders were a superconducting linear accelerator at ANL and a two-superconducting-cyclotron system at MSU. The presentations were followed by a discussion on accelerator selection. Parkinson played a major role in championing the two-superconducting-cyclotron system at MSU, shown in figure 43, and this became the working model for a new Midwest regional facility.

Negotiating Funding

Activities leading to development of a proposal for a regional facility then passed into the hands of the MSU group. A joint letter describing the spring meetings and the efforts to "formulate plans for a major heavy-ion experimental facility" was sent to NSF in April.[107] Soon thereafter, in June, the grant for the construction of the superconducting magnet prototype was approved. With funding in hand and the knowledge that Rochester would be submitting a proposal for a tandem-plus-cyclotron system around September 15, 1975, an effort began to submit a collaboration proposal by fall 1976, a timeframe consistent with possible NSF funding schedules. A letter of intent from the Midwest Collaboration would be submitted to NSF if a proposal from Rochester or elsewhere was submitted before fall.

A questionnaire to the group firmed up certain tentative organizational details and led to a September 28, 1975, meeting at which it was decided[108] to submit a group letter to NSF. The letter,[109] signed by thirty individuals, stated:

> That planning for a facility is now sufficiently advanced that we wish to notify you of our intent to submit an official proposal requesting NSF funding assistance; we expect to submit this proposal in the summer of 1976. We the signers of this letter further indicate our intent to participate in the final planning of the facility, in the preparation of the proposal, in the management of the facility, and in the use of the facility in furthering scientific knowledge.

A facility proposal was submitted to NSF early in September 1976.[110] It proposed a Coupled Superconducting Cyclotron (CSC) facility consisting of two superconducting cyclotrons: an injector cyclotron based on the prototype superconducting magnet that had been funded by the NSF and a larger cyclotron with $K = 1200$ bending power. (For complex reasons, this device was originally called a K800 cyclotron). The maximum energy would be 200 MeV/nucleon for light ions with $N = Z$, and around 30 MeV/nucleon for uranium, sufficient to study qualitatively new aspects of

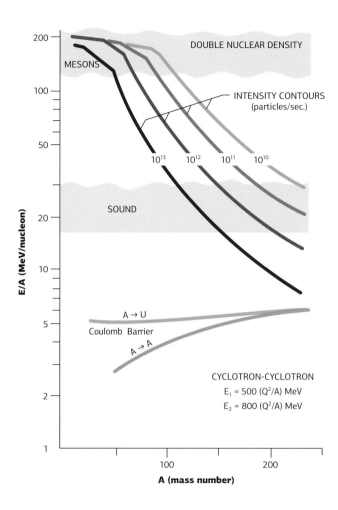

FIGURE 44. Ion energies for the proposed coupled cyclotrons. It was anticipated that the operating range would encompass two qualitative thresholds. At around 20 MeV/nucleon there would be nuclear compressional waves, or sound waves. And above perhaps 140 MeV/nucleon the colliding nuclei would produce regions with double the normal density of nuclei (200 trillion times the density of water). As shown by the curves at the bottom of the diagram, the energies are, for all masses, far above the Coulomb Barriers that prevent two charged nuclei from interacting.

nuclear phenomena as noted in figure 44. A list of forty-eight scientific sponsors from nineteen institutions—the Midwest Collaboration—was attached. The proposal included a large building addition and a significant array of experimental equipment, the two largest items being the data processing system and a large spectrograph, each costing $1,200,000. The total request was for $13,195,000 ($54,000,000 in 2013 dollars).

MSU proposed[111] to operate the facility as a national user facility, with access open to all, based on scientific merit, and with "operational management to proceed under the guidance of a national user's group." A group of sponsoring scientists would fix general laboratory policies and participate in the selection of the laboratory director, who would be advised by two external committees: a scientific advisory committee and an experiment selection committee. The director could come from outside MSU. Additional commitments of the university, during the operating life of the CSC were: to make available the present MSU Cyclotron Laboratory and its site, to provide utilities for operating the facility at cost, and to reserve up to ten housing units for visitors. If MSU became the operating organization, it would, in addition, maintain the present level of nuclear science faculty for fifteen years, provide an additional senior professorship in heavy-ion science, and continue the present MSU level of operating support for the facility staff as long as NSF's support of the facility continued.

The proposal received mostly "excellent" reviews. The less enthusiastic reviews reflected a feeling that no superconducting cyclotron had yet worked, and it would be wise to see whether a cyclotron based on the K500 prototype magnet was successful before funding such a large facility. Based on these opinions and, difficulties in finding the funds, NSF rejected the proposal. MSU reacted by submitting a revised proposal for a strongly truncated facility, Phase I, with a revised budget of somewhat over $1,200,000.

This was marginally sufficient to complete a cyclotron based on the prototype K500 magnet and to connect the K500 to existing laboratory equipment. This "bird in the hand is worth two in the bush" philosophy led to future advantages in funding. With one cyclotron already funded, it was a smaller step to complete a two-cyclotron facility. It did carry with it side effects. The K500, built on a shoestring, involved compromises in construction and would not be a reliable device until it was upgraded during a later phase of the laboratory.

The Phase I facility would still be run as a national user facility. The new proposal was reviewed, and funding was approved. It was received in three installments, a total of $1,430,000 by December 1980. With funding in hand, conversion of the prototype magnet to a K500 cyclotron began.

The NSAC Process and Phase II

A bureaucratic development on the federal level a year later turned out to be crucial for the future of the Cyclotron Laboratory. In October 1977, the NSF and ERDA (now DOE) formed the Nuclear Science Advisory Committee (NSAC, initially known as NUSAC[112]), to advise the agencies on priorities in nuclear science. It was chaired by W. A. Fowler of Caltech during 1978 and 1979. Early in 1978, NSAC was charged by the agencies to evaluate nine facility proposals that were under consideration and to present their recommendations by April 15, 1978; approved proposals would be considered for inclusion in the FY-1980 ERDA budget. An NSAC subcommittee was formed to visit some of the proposer's laboratories and to evaluate the various proposals.

NSF asked MSU to prepare a proposal, quickly, for consideration by the NSAC subcommittee. Fortunately, the Coupled Superconducting Cyclotron proposal was available and was nearly up-to-date, needing only a description of work on the previously funded K500 cyclotron and technical changes in K1200 design. A group from Milan, Italy, headed by Francesco Resmini, was responsible for most of the K1200 design and later for a similar cyclotron built at the University of Milan.[113] The Italian group worked in an office whose door was always closed, to prevent their intense cloud of cigarette smoke from invading the laboratory.

The 1976 proposal was updated by inserting addenda to the original proposal—about seventy-five new pages in all. This revised proposal[114] was submitted to NSF and the NSAC subcommittee in February 1978. It requested a total of $18,900,000 ($67,700,000 in 2013 dollars), of which $13,500,000 was for construction costs and the remainder for engineering and development. The proposal reflected the approved NSF funding for the K500 Cyclotron. It was now "for a K = 1200 post-accelerator for the K500," and "for expanded experimental areas and new experimental equipment appropriate for a research program matched to the capabilities of the coupled cyclotron system."

FIGURE 45. Francesco Resmini, who led the Italian group at MSU. He came to MSU initially as a visiting Professor in February 1978 and then was appointed as a tenured Professor in the MSU Department of Physics, and Associate Director of the Heavy-Ion Laboratory.

■ An Important Visit

From there on the procedure was rapid. Members of the NSAC subcommittee visited MSU on March 9, and from March 13 to 15 they met at the NSF in Washington. Several members of the Cyclotron Laboratory attended the meetings, including Henry Blosser and Sam Austin as Director and Associate Director, and Francesco Resmini as technical expert. Each of the nine proposers was given an opportunity to present and defend their proposals during the first two days.

Four of the proposals, from MSU, ORNL, BNL, and the University of Rochester, were for large heavy-ion facilities. The Rochester proposal was based on the use of a K = 1200 cyclotron, to be built at MSU. All except for MSU used a tandem injector followed by a large superconducting cyclotron. When the March 13 and 14 meetings were over, the MSU contingent felt, based on the various presentations and the subcommittee's reactions, that MSU would likely be the choice of these four designs, and that any competition would be with the other proposals.

The March 15 meeting was devoted to closed subcommittee deliberations, and there were further closed meetings on March 27 and 28. The subcommittee sent its conclusions to NSAC for consideration at a meeting held April 7 and 8. Fowler transmitted the NSAC recommendation—that the MSU Coupled Superconducting Cyclotron facility have the highest priority for new construction—to the federal agencies on April 14, 1978.

Smaller recommended projects involved construction of an electron-beam recirculator at the MIT-Bates linear electron accelerator and an experiment staging area at LAMPF in Los Alamos. As had been anticipated by the MSU group, the remaining heavy-ion proposals were judged either

THE NSAC COMPETITION

- Argonne National Laboratory: ATLAS, A Precision Heavy-Ion Accelerator ($4.7M)
- Brookhaven National Laboratory: A Superconducting Cyclotron Addition to the Brookhaven Three-Stage Tandem ($11.5M)
- Los Alamos Scientific Laboratory: Staging Area and Office Building ($6.0M)
- MIT: An Energy-Doubling Recirculator for the Bates Linear Accelerator ($1.7M)
- MSU: A National Facility for Research with Heavy Ions Using Coupled Superconducting Cyclotrons ($18.9M)
- ORNL: Holifield Heavy-Ion Post Accelerator ($8.1M)
- University of Rochester: A Heavy-Ion Post Accelerator ($8.1M)
- University of Washington: Post Accelerator Construction and Installation ($5.1M)
- Yale University: Upgrade MP Tandem ($3.8M)

too preliminary for approval or, in the case of Rochester, of relying too heavily on non-local (i.e., MSU) personnel. In the case of Rochester, Resmini's comments during their presentation showed that they had not yet fully mastered the MSU-based accelerator programs they employed.

The report that led to these recommendations was the first of five NSAC construction recommendations submitted between 1978 and 1982. In sum, these had the effect of developing a powerful suite of U.S. accelerator facilities. In later years, continuing to the present, NSAC produced long-range plans at roughly five-year intervals that, mostly, superseded these earlier ad hoc exercises. The first of the NSAC Long Range Plans was published in December 1979. In that plan, the MSU CSC facility was regarded as a given, already funded.

The Long Range Plan process grew out of several previous reports. One of these, the Friedlander Report (*The Future of Nuclear Science*, 1976)[115] had laid out priorities for new construction. Its highest priority was a large heavy-ion facility, and this may have influenced the NSAC Committees.

In later years, as will become clear, these five-year plans, and MSU's reaction to them, played a crucial role in the evolution of the laboratory, and ultimately to the award of the FRIB project.

■ Changing Partners

In early November 1978, there was an unanticipated and unwelcome proposal, originating in the Office of Management and Budget (OMB), to transfer the new construction project (now called Phase II) from NSF to DOE. NSF would then assume operating costs for the completed facility. This idea and the push for its implementation came from Doug Pewitt, a particle physics PhD from Florida State University who was then at OMB. His stated purpose was to use the established field office structure of the DOE to provide needed supervision for the relatively large conventional construction component of the Phase II program.[116]

Pewitt was apparently concerned about the ability of NSF to supervise a project of the size of the CSC facility and believed that it would be more likely to achieve Congressional approval and succeed in its goals if managed by DOE. A meeting of the principals—NSF, DOE, and MSU—in Washington was held on November 15, 1978. Neither NSF nor DOE was happy with the proposed change, but as it was strongly supported by OMB, it became clear that there was no way to avoid it. However, living with it would not be easy.

MSU was requested to prepare a DOE Form 44, giving all project cost details, schedules, etc., and a "Conceptual Design Report (CDR)" prepared by cutting and pasting from the CSC proposal. It was to be done within a week, if possible, with the Form 44 having priority. There would be a later meeting in DOE's Chicago Operations Office so the Chicago Office of DOE and MSU representatives could discuss how MSU and DOE would bring the project to completion, if it was funded by Congress. MSU was instructed that the documents should state then-year costs (costs in the year expenditures took place), should include adequate funding for unanticipated costs, and advised that it was important not to overstate capabilities. This quickly made it clear to MSU that learning the DOE culture and interacting with the Chicago Office would require new MSU procedures and a significant effort.

The CDR[117] was submitted on December 14, 1978, and a revised Form 44 on April 15, 1979. The contract with DOE was signed on January 20, 1980. It contained as a contractual obligation the MSU commitments that MSU President Clifton R. Wharton, Jr. had agreed to when the CSC proposal was submitted in 1976 and 1978. All in all, meeting DOE requirements caused a significant delay, but funding seemed likely to arrive. The first DOE grant was awarded on January 22, 1980.[118] Three years later, on October 1, 1983, the funding was shifted back to NSF,[119] prior to the completion of the project.

It appears that the motivation for this early funding change resulted from the continuing mutual difficulty of DOE/MSU interactions, especially concerning the delays of funding that were extending the duration and costs of the project. Certainly Blosser wished to avoid long-term interactions with certain DOE personnel, and in a note-to-self,[120] he considered whether DOE personnel could be changed and said, "I can't accept project as it's presently formulated." These feelings were apparently reciprocated.[121]

Making It All Work

Following the award of the Phase II project in 1978, it was clear that the MSU Cyclotron Laboratory faced a daunting set of tasks. The first was to complete the Phase I system: NSF had funded the K500 cyclotron and an array of experimental equipment that would serve until the Phase II system came on line in 1984. The second was to prepare for the arrival of funding for the Phase II facility and the inevitable conflicts with Phase I as Phase II construction began. The laboratory had also to organize itself as a national user facility.

Of course, during all this time the laboratory had been running a forefront research program on the K50. April 1, 1979, had been set as the date to close the K50 and concentrate on K500 installation and operation. This date was delayed slightly, to July 1, 1979, so as to finish a few graduate student theses.

More than 175 people attended a formal K50 decommissioning ceremony that was held on September 9, 1979, with talks by: Larry Kull, the first PhD to do a thesis on the cyclotron; Dr. Milton L. Muelder, MSU Vice President for Research during K50 funding; and Dr. William Rodney, Program Director for Nuclear Physics at the National Science Foundation. Rodney was associated with the K50 during its entire history.

■ Phase I and K500 Cyclotron Development

When funding was received in mid-1977,[122] it was expected that the new K500 cyclotron would be operating by early 1980. Many of us thought this schedule was optimistic: it assumed that the new cyclotron, involving many new technologies and competing with the Phase II program, would be complete in only two and a half years. And many unanticipated difficulties in the Phase II construction process also delayed K500 completion. In the end, external beam from the K500 was obtained on August 31, 1982.

The major K500 cyclotron-related tasks were developing the radio frequency and beam extraction systems. In addition were the many items, each relatively small in itself but large in aggregate, required for cooling water, magnet power, control of the many subsystems, and providing radiation safety. A beam transport system had to be built, as well as two new spectrometers, a large scattering chamber, and a new data handling system, with new computers. Building and operating the K500's radio frequency system, however, proved to be the greatest challenge.

The K500 magnet was moved into a new MSU-funded building addition in March 1978 and operation resumed in May 1978. Its first use was to test ion sources for heavy ions and see if they worked equally well at these high magnetic fields. They did.

After the ion-source tests were complete, the prototype cyclotron magnet was completely disassembled to install the extraction elements and the field-trimming coils that precisely shaped the magnetic field. After reassembly, the magnetic field was measured in detail and the effects of the trim coils were determined. When the magnetic field measurements were finished in November 1980, the magnet was again disassembled to install the radio frequency system.

The following reassembly was difficult and took until mid-1981. Part of the difficulty owed to the compactness of the high field cyclotron: both the central region of the machine and the extraction system involved tight tolerances and needed precise construction and simulation. Finally, at the end of September 1981, the K500 was ready for internal beam tests although some hardware, including that for extracting the beam from the cyclotron, was still missing.

It was found that operating the K500 with the originally proposed radio frequency range would be difficult, especially for injection into the K1200 cyclotron. That problem was solved by changing to lower frequencies, nine to thirty-two MHz (one MHz is one million cycles per second; FM frequencies are about 100 MHz). This, however, required the introduction of insulators into the

FIGURE 46. K500 cyclotron central region. The bottom hills are shown, and there is a mirror image top hill about one inch above each bottom hill. The top and bottom parts of the dees are also shown. The beam circulates through a space between the top and bottom parts of the dees that is at the same height as the space between the top and bottom hills. The two parts of the dee are supported separately, but only the top supports (dee stems) are shown.

long tubes, the dee-stems, that supported the three dees from above and below as shown in figure 46. In October 1981, several of the large dee-stem insulators cracked and there was a shortage of the highly specialized vacuum seals for their replacements. It was also difficult to obtain the proper balance and timing of the radio frequency voltages that were applied separately to each of the three dees.

Many of these problems became apparent, and crucial, when first attempts to achieve internal beam were made in November 1981.[123] First tests with $^{12}C^{2+}$ ions on November 6–8 suffered from unbalanced, out-of-phase dee voltages that could only be partially compensated. November 13 tests with the same ions were somewhat more successful, but the poor vacuum prevented the beam from being taken to the full radius. After substituting deuterons, which were less sensitive to poor vacuum, for ^{12}C, the beam was taken to full internal radius on November 24, 1981.[124] It would be ten months before it was possible to extract the internal beams from the cyclotron.

After the achievement of internal beam, a number of improvements were made: a fix for the radio frequency problems, followed by installation of the final vacuum system; electrostatic deflectors, a crucial part of the extraction system; and a beam probe that gave detailed information on the internal beam. Finally, on August 31, 1982, a beam of 106 MeV deuterons was extracted from the K500; beams of 215 MeV alpha particles and 420 MeV ^{12}C ions soon followed. Then we celebrated (see fig. 47).

After a gestation period two years longer than anticipated, the K500 was finally ready to operate, and the suite of detection apparatus was nearing completion. These beams provided the first test of the new beam transport system and were used in early experiments, mostly in a sixty-inch-diameter scattering chamber. Although some initial experiments on an incomplete S320 were done in 1983, the full array of Phase I detection equipment, in particular the S320 and RPMS spectrometers, was not completed until mid-1984.

During this process we lived with a potentially serious problem that turned out to be unimportant: the K500 coil had an electrical short. A short in a superconducting coil is not necessarily disastrous because a short always has a finite, if small, resistance. Therefore, once the wire is superconducting, with zero resistance, a short will not usually be noticed, except, possibly, when one is increasing the coil current. The rate at which one can do so could be limited. Of course, the problem might worsen, with a failure occurring at an inopportune time.

As became common practice later in the lab history, a review team was assembled in March 1981, to examine the problem and make a recommendation. Perhaps influenced by the availability

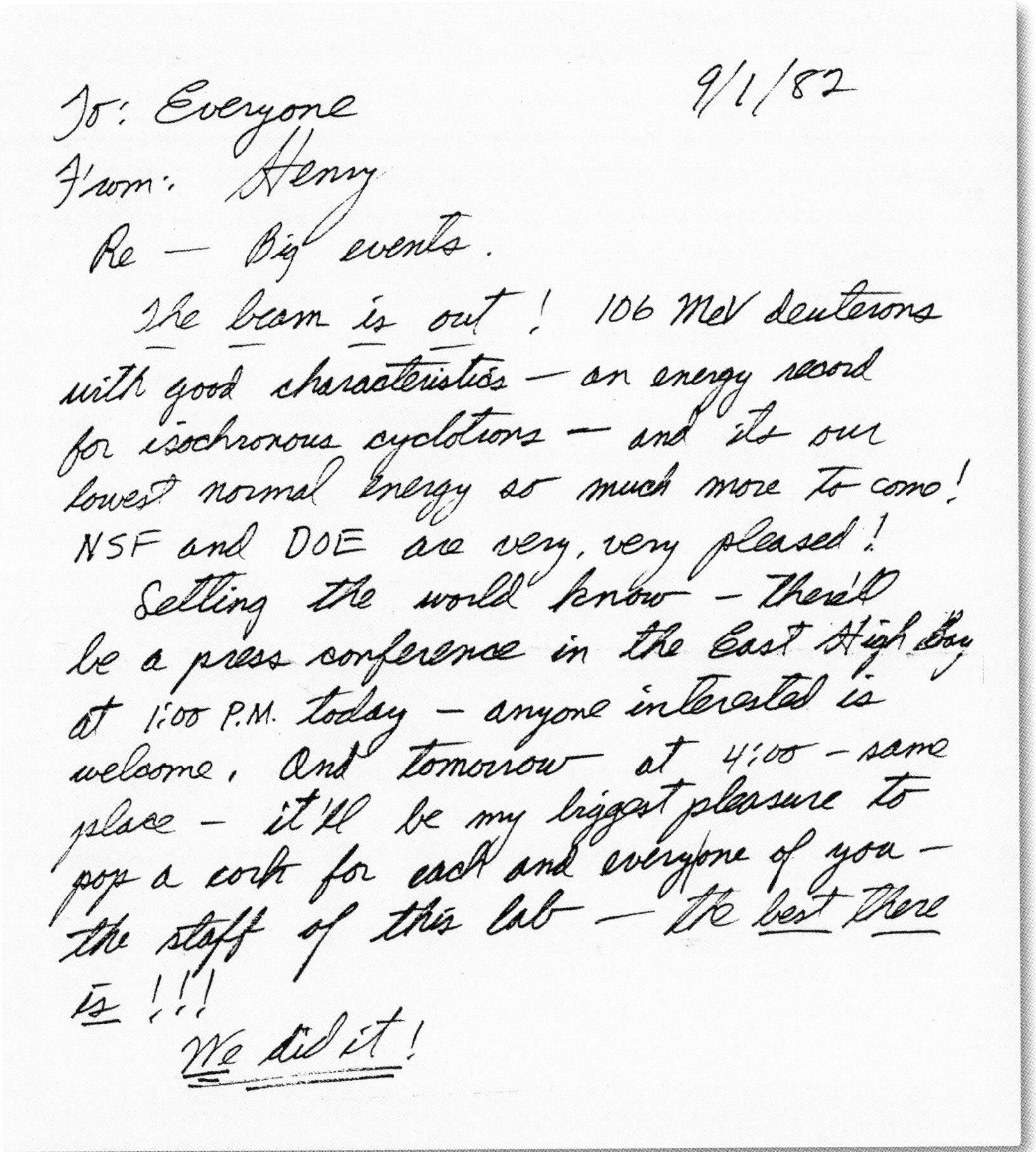

To: Everyone   9/1/82
From: Henry
    Re — Big events.
        The beam is out! 106 MeV deuterons
with good characteristics — an energy record
for isochronous cyclotrons — and it's our
lowest normal energy so much more to come!
NSF and DOE are very, very pleased!
        Letting the world know — there'll
be a press conference in the East High Bay
at 1:00 P.M. today — anyone interested is
welcome. And tomorrow at 4:00 — same
place — it'll be my biggest pleasure to
pop a cork for each and everyone of you —
the staff of this lab — the best there
is !!!
        We did it!

FIGURE 47. Blosser's handwritten note announcing this success and inviting the lab staff to what was a typical celebration of such successes: a champagne fest. This might not have been in full accord with MSU regulations.

of the spare superconducting wire, the team's recommendation was to proceed with winding a replacement coil. However, because of the press of Phase II construction, the probability of eventual failure had to be balanced against the significant time, money, and effort that would be spent in rewinding and testing the coil. In the end, winding was postponed, then postponed again, and was never done. The flawed coil has worked without incident for over thirty years.

The Cyclotron Laboratory has often used such external review teams, both to respond to problems and to provide information before proceeding on large projects. This practice has had a large positive benefit for the laboratory. Most important is that the additional information obtained while planning for a review and from the review itself helps avoid poor decisions. Recommendations of such review teams, however, tend to be conservative, since no one wishes to overlook or minimize a problem, so one has to balance the cost of carrying out recommendations in marginal cases against the danger of doing nothing. But even if nothing is done, the reviews were a benefit because many influential nuclear scientists came to know the laboratory and its internal expertise, and were usually impressed.

■ Becoming a National User Facility: The NSCL

In the middle of this period of intense work, the MSU Cyclotron Laboratory was no more. On October 1, 1977, once MSU received funds to complete the K500 cyclotron, the MSU Board of Trustees approved a name change to the "MSU/NSF Heavy Ion Laboratory."[125] It was officially renamed again on January 22, 1980, as the National Superconducting Cyclotron Laboratory (NSCL). Although this name was chosen by the DOE, NSCL is universally used for this NSF National User Facility.

Becoming a national user facility carries with it a number of requirements, some mainly external and some mainly internal. These include:

- A USER GROUP. Open to anyone, primarily nuclear scientists, who are interested in the NSCL or its activities. This group has grown over the years, and as of summer 2015 has over 1,500 members from about fifty countries.[126]
- A USERS EXECUTIVE COMMITTEE. A committee is elected by the User Group to convey user concerns and comments to the NSCL Director and to the federal agencies. Members serve

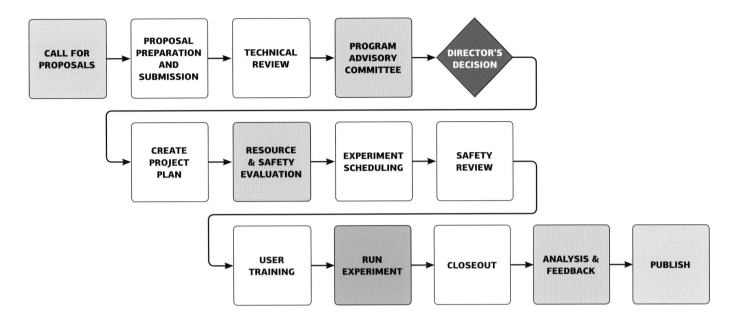

FIGURE 48. The Program Advisory Committee process, followed by the steps of preparing for and running an experiment, and then making its results public.

two-year terms (later extended to three years) and elect their own chair. (See Appendix E for a listing of members.)

- A PROGRAM ADVISORY COMMITTEE. NSCL beam time is granted based purely on scientific merit without preference for local scientists, following the process outlined in figure 48. For this purpose a Program Advisory Committee is appointed by the NSCL. Its members (initially six members, now eight) usually serve staggered three-year terms. The PAC advises the Director about which experiments should receive beam time. The PAC does not have a chair but is convened by the Director, or recently, by the NSCL Chief Scientist; the convener does not have a vote. The first K500 PAC met in February 1982 to consider proposals for beam time submitted in response to a call for proposals. Forty-three proposals for a total of 4,334 beam hours were received and thirty-four experiments were approved for 1,586 beam hours. As of June 2015, there have been thirty-nine PAC meetings for this purpose.
- THE BEAM-TIME APPROVAL PROCESS. Users typically have eight weeks to respond to a call for proposals. They are informed of the individual(s) responsible for an in-depth review of their

proposals and may contact them with further information if they choose. The PAC meets some weeks later; considers the comments of the in-depth reviewer and their own judgment of the proposal; makes recommendations of who shall receive beam time; and provides the NSCL Director with a priority recommendation to assist in scheduling beam time. The results are announced immediately. Approvals are valid for twenty-four months after the decision is reached. After that a proposal must be resubmitted for consideration by the PAC. If developments in the field make a rapid response desirable, experimenters may request discretionary time from the Director; such discretionary grants are limited. For a list of PAC members, see Appendix E.

- A DIRECTOR. For the NSCL, the Director is chosen by the MSU administration, with the advice of the Users Executive Committee and the federal agencies. The Director could be from another institution, but although external candidates for director have been considered, the two directors following Blosser, Sam Austin and Konrad Gelbke, have been from MSU. This procedure has changed in the FRIB era. For a list of Directors and Associate Directors (chosen by the Director, with the concurrence, and in the earlier days, the vote of the laboratory faculty), see Appendix C.

- ENHANCED FACILITY OPERATIONS. A facility open to all users must have cyclotron operators. A large, high-power, high-tech device is fragile, and its operation requires special expertise; users from other institutions cannot be expected to learn these skills. An initial group of operators was hired and trained with NSCL, especially in the operation of complex beam line equipment. This was a major change for NSCL laboratory members who had used the K50.

During the K50 era, everything, including beam production and tuning, had been done by the experimenters. Operation now evolved toward a new system where beam production became a "black-box" (see fig. 49), with experimenter's responsibility beginning at the experimental apparatus. But even so, facility operations were difficult and unreliable at first during this transition, and lab faculty often had to help with the operation process, especially for users from outside NSCL.

This evolution of responsibility is part of a continuing process. In the early days of physics, experimenters built their own equipment including, for example, vacuum tubes and vacuum pumps. Increased specialization became inevitable as experiments became more complex.

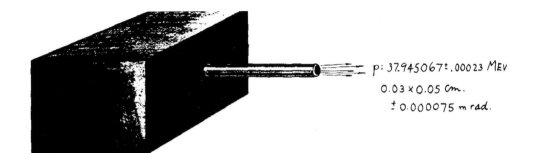

$p: 37.945067 \pm .00023$ MEV
0.03×0.05 Cm.
± 0.000075 m rad.

FIGURE 49. This "black box" approach is captured by this David Judd cartoon: "The Cyclotron as Seen by the Experimental Physicist."

■ Evolution in Laboratory Management

In addition to changes related to becoming a national user facility, there were many other changes, some related to commitments made to DOE, and others initiated to further the efficiency of construction and operation.

The MSU commitments to DOE, as carried over to NSF, included right of consultation on the following:

- choosing the Director;
- making available the MSU Cyclotron and its site for agency purposes;
- maintaining the present level of faculty involvement in nuclear science for at least fifteen years (later interpreted[127] to extend to June 30, 1995, fifteen years after the signing of the DOE contract);
- providing an additional professorship in heavy-ion science at the most senior level; and
- maintaining partial operating support of the laboratory at the then-current level.

These commitments proved important through the early part of the Phase II project when success was not assured, and there were temptations to use part of the site for other purposes and to convert nuclear physics faculty vacancies to other research areas in the Physics Department. Later, the NSCL was easily able to defend itself based on performance, and the personnel commitments became a source of paperwork with few compensating advantages.

Effective August 1, 1979, the Director of the NSCL became a Separately Reporting Director within the MSU Administrative structure, responsible to the MSU Provost and holding a rank equivalent to Dean. Henry Blosser was the first NSCL Director, and negotiated a number of agreements with the MSU administration that made the NSCL unique among MSU units. For example, NSCL

- handled its own purchasing actions, with only after-the-fact review retained by the university units—the NSCL began issuing its own purchase orders on March 12, 1980;[128]
- received a continuing contribution to the NSCL of a relatively small base budget, to be incremented by salary raises;
- received consideration of research and development funds for laboratory improvements that could be substantial; and
- instituted (in 1980–81) a Continuing Appointment (CA) system to provide individuals with contract tenure (analogous to that at National Laboratories and some other universities) and help to attract high-level employees.

As noted previously, during its early years, the Cyclotron Laboratory had an extremely flat administrative structure: a director, for most of this period Henry Blosser, his administrative assistants, and members of the faculty who were informally consulted on important matters at the "Blosser-house" meetings. Faculty often had ad hoc administrative duties, being in charge of beam line or apparatus construction, for example.

However, as the laboratory undertook a large construction project, the number of personnel (faculty, postdoctoral, student, and technical) greatly increased, and it became clear that the structure had to evolve. The first step in this direction had been instituting the office of Associate Director in 1975; the first AD was Hobson Wildenthal. The AD assumed most administrative duties, excluding those associated with accelerator operation/construction and relations with the NSF and the MSU administration. The Director/Associate Director relationship was much like that of Dean/Department Chair. The AD handled student and postdoctoral appointments, space assignments, faculty procurement requests, cyclotron scheduling, negotiations with the Physics Department on division of resources and on salaries, etc. The precise division of duties depended somewhat on who was director and also evolved with time. The AD appointments are listed in Appendix C.

From 1983 to 1985, this position was temporarily replaced by a Research Director, Sam Austin,

FIGURE 50 (*opposite*). Co-directors Sam Austin (*left*) and Henry Blosser in 1987, two years after the co-director arrangement began. They are standing in front of the superstructure of the K1200 cyclotron.

with essentially the same duties, but also charged with representing both inside and outside users of the facility on an equal basis as far as cyclotron access and use was concerned. In addition, the RD was in charge of K500 operations. Given the unreliability of the K500, this was a time-consuming task.

Then, in 1985, when the construction of Phase II was seriously behind schedule, the NSCL, with the permission of the NSF and MSU, instituted a co-director arrangement, where Henry Blosser was in charge of Phase II construction and Sam Austin of all other laboratory matters, including interactions with the NSF and MSU administration (see fig. 50). The Associate Director position was reinstituted and the Research Director position was eliminated at this time.

Certain aspects of these changes involved the Physics and Astronomy Department (PA); the Physics Department had been combined with Astronomy during a budget crunch in 1981.[129] The most important issues were arrangement (purchase) of a reduced teaching commitment for an NSCL Associate Director performing NSCL administrative duties, the commitment to maintain the size of the nuclear physics faculty for fifteen years, and PA joint faculty appointments with the NSCL to monitor these commitments. During 1975, these commitments resulted, once again, in conflict; some PA members believed they allowed NSCL members to avoid teaching duties and increase those of the other PA faculty. Eventually these issues were settled, partly owing to the formal commitments made by the MSU administration to DOE as described earlier in connection with the CSC facility.

The K500 Experimental Program[130]

The first K500 experiment used a ^{12}C beam with an energy of 35 MeV/nucleon and began shortly after the first beam was extracted in September 1982. It was performed by Gary Westfall and his collaborators,[131] which included everyone with an interest in heavy ion–induced reactions. As can be seen from figure 51, this was the beginning of a new sort of science for the laboratory, where one is interested in the properties of nuclear material and not in the positions and properties of individual quantum states. Initially, most of the accelerator time was devoted to developing different beams, and by March 1983, beams of ^{4}He, ^{12}C, ^{16}O, ^{20,22}Ne were available at energies, depending on the isotope, of between 15 and 35 MeV/nucleon.

By February 1983, the facility was sufficiently reliable that the first experiments involving outside users were run, although schedules were still hard to maintain. When the first experimental period ended in July 1983, fourteen approved experiments had received beam time. Most used relatively light ions and were done in the sixty-inch scattering chamber.

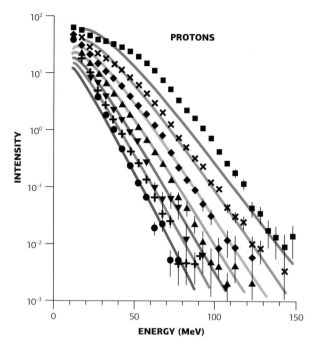

FIGURE 51. Spectra of protons at different observation angles produced by bombardment of a gold target (A = 179, Z = 79) by 35 MeV/nucleon ^{12}C ions. Proton energy is on the abscissa and proton intensity on the ordinate. These spectra were compared to quantum statistical, hydrodynamic, and thermal models. The thermal model assumes that a moving source at temperature T emits particles. One finds T is about 10 MeV, which in more familiar units is 100 billion Kelvin or 100 billion degrees centigrade (Celsius). These are perhaps the highest temperatures seen in nature.

■ Start-Up Challenges

Although we had made a good start, we still needed to improve the reliability of the K500, and to finish development of the new spectrograph, the S320 shown in figure 52, and the Reaction Products Mass Separator, the RPMS. Jerry Nolen was the principal architect of both of these devices.

We needed a new spectrograph because the Enge split-pole magnet was not powerful enough for many of the beams from the K500. Funds, however, were extremely limited and the major magnets for the S320 had to be obtained secondhand from other laboratories. While the S320 had relatively low resolution, it was adequate for most K500 experiments; the Cyclotron Laboratory no longer did high-resolution experiments. The S320 also served as a switching magnet to guide the beam to other experiments.[132]

The RPMS was a more specialized device, built to accept high-speed reaction products emerging

from the target, bend them according to their mass-to-charge ratio, and focus them to points on a focal plane.[133] Their decay could then be studied.

The RPMS was also mainly built from secondhand parts. Its major component, a large Wien filter that sorted particles according to their speed, was obtained from the Lawrence Berkeley Laboratory. A particle-bending magnet was obtained from Cornell University, and quadrupole focusing magnets from Cornell via Maryland, and from SREL. Eventually, the RPMS was superseded by better technology—the fragment separators we shall encounter later.

Many experiments were also done in a sixty-inch diameter scattering chamber shown in figure 54.

At the end of July 1983, the K500 was shut down for a variety of repairs and improvements: fourteen items were on the list; an extensive remapping of the magnetic field to better identify

FIGURE 52. The S320 spectrograph. The main bending magnet, on the far left, was a beam bending magnet from the University of Pennsylvania tandem accelerator laboratory, and the main quadrupole magnet was from the Space Radiation Effects Laboratory (SREL) in Newport News, Virginia. Both of these laboratories had been closed.

FIGURE 53. Brad Sherrill near the S800 spectrograph. Sherrill received his PhD from Michigan State University in 1985. After a year at GSI, he returned to MSU and was promoted to University Distinguished Professor of Physics in 2003. At the NSCL and FRIB he has played a major role in the development of fragment separators and spectrographs and in pioneering experiments with rare isotope beams. He was Chair of the Division of Nuclear Physics of the APS in 2005, and served as FRIB/NSCL chief scientist from 2009–14. In 2015 he was named NSCL Director.

undesired first harmonic components of the field had the highest priority. An oversight during these repairs allowed a water line inside the dees and the dee stems to freeze and resulted in copper tubing ruptures that added a month to the shutdown and led to future maintenance problems. Finally, on December 20, 1984, we began running the experiments approved earlier by PAC 1 and PAC 2.

K500 operation was now somewhat more reliable: one could plan for seventy-five to eighty hours per week (3,000–4,000 hours/year) of beam time. New beams of lithium isotopes, carbon isotopes, and argon isotopes had become available, as had higher energies, typically 35 MeV/nucleon for the heavier ions. The fraction of the internal beam that could be extracted improved from 35 percent to 50 percent. Much of the PAC 1 beam time was devoted to the development of experimental apparatus, so that by July 1984, all the Phase I apparatus (S320 Spectrograph, Reaction Products

FIGURE 54. Experiments in the sixty-inch scattering chamber. With the K50, high resolution was the goal. With the K500 accelerating heavy ions, one was satisfied with 1 to 2 percent energy resolution, which could be achieved by simpler detectors, capable of identifying the type of particle (p, or alpha, for example). One could observe ten or so angles simultaneously. The target is a thin foil mounted near the circle in the center-bottom of the chamber. The detectors are the tubular devices aimed at the target.

Mass Separator [RPMS], sixty-inch scattering chamber, Enge split-pole spectrograph, and a general purpose gamma-ray chamber) was complete and had been used in experiments. Most of the PAC 2 experiments were complete by then: PAC 3 met on July 2, 1984, and PAC 4 on January 13–14, 1985. From then on PACs met about every six months. The most heavily used instruments were the S320 Spectrograph and the sixty-inch scattering chamber.

■ Exploring Ion-Source Solutions

Up to now, the laboratory had been using Penning ion sources,[134] a long-term standard in cyclotrons, which were convenient because they were small and were located at the center of the cyclotron. However, they suffered from substantial limitations when used for heavy ions. They had short lifetimes: between two and twenty hours, depending on the support gas and the ion species,[135] and produced ions with relatively low charges. This meant that experimenters were continually changing ion sources, a major handicap for efficient operation. The negative effects of the gas load on the vacuum and radio frequency systems were another big drawback of these internal sources.

We had been considering building an Electron Cyclotron Resonance (ECR) ion source[136] for use with both the K500 and K1200 cyclotrons. An ECR source is based on a gaseous discharge, roughly similar to that in a neon tube, powered by a radio frequency generator. The gaseous discharge strips electrons from the atoms making up the gas, the resulting plasma is confined by a complicated arrangement of magnetic fields, and the ionized atoms are then pulled out of the discharge by electric fields. It can produce ions with large charges, for example, a carbon nucleus with all its electrons removed. ECR sources had great potential: long lifetimes—weeks to months compared to hours for our Penning (PIG) ion sources—and production of higher charge states. Since energy from a cyclotron is proportional to the square of the charge, this would be a great advantage.

An external review team examined our ECR plans in October 1984. This review led to a decision to build a room temperature source about 30 percent larger than a working source at LBL. If performance increased with size as hoped, a still larger ECR with superconducting coils was planned. This decision would have a great influence on the direction of Cyclotron Laboratory development.

A plasma discharge was observed for the first time in the RT-ECR (RT for Room Temperature)

FIGURE 55. The Room Temperature Electron Cyclotron Resonance Ion Source (RT-ECR) with its builder Tim Antaya. The source contains a complex set of magnets, both circular coils and hexapole coils. Completed in August 1985, it was the laboratory's first ECR source. The many ECRs constructed later are summarized in the section on "Ion-Source Generations," on page 232. The argon gas discharge pictured above reflects the six poles of the hexapole. (The picture is from another ECR source, Artemis A, which was built later.)

on July 22, 1985, and the source shown in figure 55 was completed in August 1985.[137] Eventually the RT-ECR obtained world-leading intensities for krypton ions.[138]

The specialized magnets that confined the gaseous source plasma had to be placed outside the cyclotron, as did the radio frequency source that ionized the plasma and the shielding against the large flux of X-rays from the ECR. A beam line led the beam extracted from the ECR toward the cyclotron and bent it upward in the direction of the cyclotron's vertical symmetry axis. It was not straightforward to develop the system that completed the injection into the cyclotron.[139] The spiral inflector that bent the injected beam back from the vertical into the horizontal orbit of the K500 had a complex shape and had to be small to fit into the compact geometry of the K500. The rather large magnetic fields in the vicinity of the cyclotron were another major complication. A beam line adapted to deal with these fields was not available until the K500 upgrade in the late 1990s.

The first ECR-produced beam from the K500 was extracted on March 25, 1986. As anticipated, higher beam energies were obtained, especially for the heavier ions such as argon and krypton. Penning ion sources have not been used since that time.

We soon learned that operation with a single ECR source was inefficient. It took a long time to develop new beams, and operating with certain beams poisoned the source for others. The next planned source was a superconducting ECR, but fearing the development of such a source would be long and involve unanticipated difficulties, we instead built a small, relatively inexpensive source for alkali elements (lithium, sodium, potassium, etc.) and other metals. This Compact ECR (CP-ECR) was finished in late March 1987, and put into immediate use, thereby freeing up time for development of beams in the RT-ECR. Later the CP-ECR was adapted for low charge state gaseous ions and used as the injector source for the K1200 for tests and relatively undemanding beams.

During the K500 era, experiments studying the structure of nuclei and those studying the nature of nuclear reactions—especially collisions of a heavy-beam ion with a heavy-target nucleus—were equally common. With the development of the ECR source, reaction experiments with higher energy and heavier ions became possible. Such experiments greatly benefitted from detectors that could simultaneously observe reaction products produced at all angles. The first of these detectors—the 4π Array, the Washington University Dwarf Ball-Wall array, and the MSU Miniball—see figures 69 and 70—were used in some of the last K500 experiments.

Frequent breakdowns still made it difficult to run an efficient experimental program on the K500. Some problems arose from unfamiliarity with the construction and operation of a unique

FIGURE 56 (*opposite*). A comparison of two K = 500 cyclotrons: the TRIUMF cyclotron shown here has a diameter of fifty-nine feet, and the MSU K500, shown as a red circle, has a diameter of ten feet. The TRIUMF cyclotron's magnetic field is roughly a tenth that of the K500, which means it must be much larger to have the same bending power. This is, however, to some extent a comparison of apples and oranges. The TRIUMF cyclotron accelerates negative hydrogen ions to 500 MeV, which strongly limits the magnetic field that can be used, if the accelerated ions are to survive.

type of machine, but many could be traced to the compromises made to build the K500 within a marginal budget. In addition, experimenters desired a great variety of beams, and the need to develop them prevented standardization. The breakdowns usually consumed 30 to 40 percent of the scheduled time: from 1985 through 1990, the breakdown rates were 45, 25, 29, 15, 43, and 43 percent, respectively.

Since outside users had usually spent significant funds travelling to MSU/NSCL, we alternately scheduled inside-user and outside-user experiments and let the inside schedule slide in order to finish outside experiments. Often the inside experiment then had to be rescheduled at a later time.

This made it difficult to plan experimental schedules and to maintain the morale of the local experimenters on whom the efficient operation of the facility depended. A lesson learned the hard way was that one needed breakdown rates of 10 percent or less in order to run an efficient program involving external users. Obtaining such reliability governed much future lab development.

■ Compactness: Pros (Mostly) and Cons

Given the overall K500 experience, it seems appropriate to discuss the pros and cons of the extremely compact design of the superconducting cyclotrons. Compactness carries with it strong advantages: much less iron is required, and a smaller footprint requires less building space. Comparing two K500 cyclotrons in figure 56 and superconducting versus normal cyclotron masses in figure 57 emphasizes this point. Electrical power costs are also reduced because energy losses in the magnet coils, now superconducting, are eliminated. A K1200 cyclotron and its beam lines, using room-temperature technology, would be inordinately expensive to build and operate, and would probably not be affordable or fundable.

Miniaturization also carries costs. High-power beams can easily melt small components. An aversion to the cost of building thick, heavy, magnetic-flux return yokes when the rest of the cyclotron was so small also had negative effects. Return yokes such as those shown in the pill box shape of the magnet in figure 38 are necessary to confine magnetic fields within the cyclotron. With thinner yokes, magnetic fields extend far from the accelerator.

These fringe fields strongly, and negatively, affected the performance of the beam line leading from the ECR source to the K500, the operation of unshielded radio frequency amplifiers, and many

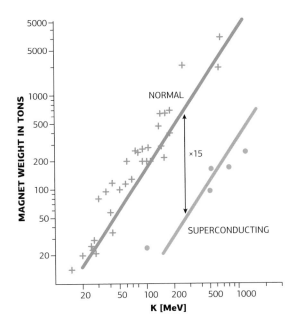

FIGURE 57. The magnet weight of normal temperature and superconducting cyclotrons. The three MSU superconducting cyclotrons with K values of 100, 500, and 1,200 MeV all lie near the lower curve. On average, for the same bending power (K), normal temperature cyclotrons weigh fifteen times as much as superconducting cyclotrons, as indicated by the two-headed arrow.

diagnostic instruments. One had to worry about wrenches and vacuum pumps being "sucked up" by the stray fields, so the lab bought bronze wrenches. Many of those who worked on the details of beam optics and hardware wished the cyclotron had been somewhat larger and had a thicker return yoke to avoid these problems.

On the other hand, the difference in magnet weight for normal temperature and superconducting of the same bending power is large. Sometimes an impressive, although somewhat unjustified, comparison was made with the TRIUMF Cyclotron, as shown in figure 56. A more realistic comparison is made in figure 57, data from W. A. Barletta, which shows the magnet weights of normal temperature and superconducting cyclotrons such as the K500. The superconducting magnets weigh far less, by a factor of about fifteen. It seemed clear that these arguments made it more likely that funding would be obtained. This approach was also useful in the design of the extremely compact cyclotron that the NSCL later built for treating cancer with neutrons at Harper Grace Hospital in Detroit.

The reliability of the compact superconducting cyclotrons does not seem to be an intrinsic problem. As we next describe Phase II, we will see that, with attention, high reliability is possible.

The Phase II Project

As noted earlier, when funding for completion of the K500 was received in 1977, it was expected that the new K500 cyclotron would be in operation in early 1980. When, shortly thereafter in 1978, NSAC recommended funding of the MSU Phase II project to begin in Fiscal Year 1980, the MSU Cyclotron Laboratory suddenly faced a complex set of potentially conflicting goals: complete the Phase I (K500) system; prepare for beginning the Phase II facility; and organize the laboratory as a new national user facility.

The 1978 NSAC recommendation for FY-1980 facility construction included the MSU Phase II project at a cost of $18.9M in 1978 dollars, with funding to begin in FY-1980 and beam projected to be available early in 1984. When the DOE contract was signed on January 22, 1980, the estimated project cost was $30M, including contingencies of $4.3M, $4.2M in research and development related to construction, and extension of the funding over five years, all of which increased the original cost estimates. Inflation was also a major factor: the consumer price index had increased by over 25 percent from 1978 to 1980.

■ New Lessons to Learn

There was optimism that construction would go more smoothly for the K1200 than for the K500 because that machine had served as a prototype, and the known errors in its design had been corrected. This optimism turned out to be only partially justified because the higher energy design was in fundamental and unanticipated ways more difficult than that of the K500. And, as with the K500, apparatus providers often did not meet their specifications.

The most time-critical item was to construct an additional building to house the K1200 cyclotron, the extended experimental area, a larger machine shop, and offices for construction personnel. With this in mind, Phase II architectural studies had started two months prior to the formal contract signing, based on a DOE authorization letter. Plans reached final form and were approved by DOE and issued for bids in September 1980.[140] The addition, essentially complete by early 1982, is labeled 1982 in figure 84. Part of it is being replaced by the 2016 addition.

Orders for the long-lead-time items were placed quickly, but their outcome was often unfortunate: of seven major procurements, only the one for the computer system was more or less problem free.

Some sections of the superconducting wire for the magnet coil did not meet the current carrying specifications, and the coil design had to be changed to allow the use of the poorer conductor in regions of the magnet where the magnetic fields were smaller. Nor was the wire of uniform cross section, which required a modification of the coil-winding apparatus.

The supplier of the bobbin for the coil also failed; the bobbin was greatly deformed in the welding process. We took over the remanufacture of the bobbin and also of the cryostat for the magnet after another manufacturer failure. And the central refrigerator helium liquefier arrived with twenty-five missed specifications.

There were also flaws in the iron for the magnet: correctable errors in mechanical tolerances and internal voids in the magnet iron. Some parts had to be recast.

Of longest-term impact, the manufacturers of the radio frequency power supply failed to make a transformer that could stand the shock of a dead short (called "a crowbar") as sometimes occurs during operation and as was required in the specifications.

All these issues, except for the transformer problem, were handled, although with some delay, and partly funded by refunds from the vendors. Going forward, we shall describe, where appropriate, how the transformer problems continued for many years.

FIGURE 58. The steel for the K500 and K1200 magnets was produced at the Bay City Foundry. Pouring steel is always a dramatic event.

■ Almost Overwhelmed

Obtaining adequate manpower for all these projects was a challenge. One particular win-win situation was that Texas A&M, which was building a K500 cyclotron based on the MSU K500 design, agreed to take over the winding of the K1200 coils in return for using MSU/NSCL facilities shown in figure 59 to build the coils for their own K500.

A more fundamental problem was extraction of the beam from the K1200.[141] It had been assumed that the computational techniques used for the K500 cyclotron could also be employed at the higher K1200 energies, that is, that the small radius and extraction radius orbits could be treated separately.

But it turned out that the magnetic elements necessary for extraction perturbed the small-radius turns in the strongly spiraled magnetic field elements required for the higher energies of the K1200. This prevented separate treatment of the small- and large-radius orbits. In addition, it was necessary to reduce the voltages on the electrostatic parts of the extraction system to be compatible with the voltages that had proved obtainable in the K500.

As a result, an iterative design procedure had to be used, requiring a massive amount of computing power. A rush proposal to NSF for a more powerful (ten to twenty times faster) computer[142] was funded in December 1984. A flurry of calculations, some on the previous computer system and more detailed versions on the new FPS 164 computer, were eventually successful. By January 1986, the design of the extraction system had been frozen. It was finally possible to construct the extraction elements and make the necessary exit ports in the K1200 for extraction of the beam.

It had become clear that, in addition to the major delay in the construction schedule, there was a partially related cost overrun in the construction budget, mainly because construction manpower had to be paid. The delays and overruns were part of the motivation for setting up the a co-directorship management arrangement mentioned earlier, to allow Henry Blosser to concentrate on accelerator construction, while other NSCL administrative duties were handled by Sam Austin.

Another motivation for creating the co-directorship was Blosser's increasing interest and involvement in conceptualizing and constructing superconducting cyclotron systems for cancer therapy. In 1981, he had met William Powers, an oncologist who had a similar entrepreneurial personality. They evolved the idea of building a neutron-producing superconducting cyclotron small enough to be mounted on a circular gantry and rotated about a patient.

Blosser spent only a small percentage of his normal working hours on this device, but it was

FIGURE 59 (opposite). Team from Texas A&M winding the K1200 coil. The coil is wound on a spool mounted on a large lathe that turns to wind the coil. The first layer of wire is being wound on the bottom segment. The round spool to the right of the coil holds insulation, which surrounds each strand of the superconducting wire.

DIRECT IMPACT: CYCLOTRONS FOR CANCER THERAPY

The NSCL, driven by Henry Blosser's interest in societal applications, pioneered applications of superconducting cyclotrons for cancer therapy. After receiving a grant from Harper Hospital in Detroit,[143] the NSCL used its experience in cyclotron miniaturization to build a very small cyclotron that could accelerate deuterons. The deuterons bombarded a beryllium target and produced neutrons for treating cancer, especially prostate cancer. The cyclotron was so small it could be mounted on a gantry and rotated around the patient under treatment. Because the facility could bombard a patient with neutrons from several angles, always aimed at a cancer tumor, the surrounding tissue received less radiation than the tumor and was less likely to be damaged.

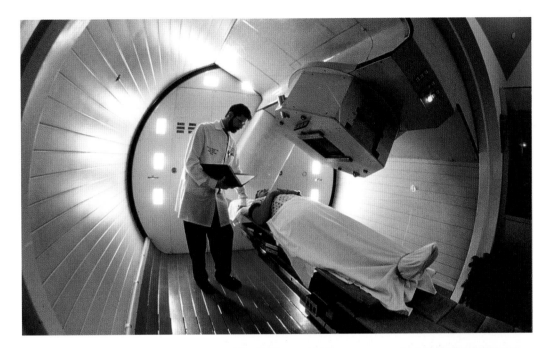

The cyclotron was installed in the Gershenson Radiation Oncology Center at Harper Hospital in Detroit on July 10, 1990, and from 1992 to 2012 treated over 2,000 patients.[144] Later, the NSCL designed[145] a cyclotron that produces beams of 250 MeV protons for cancer therapy; two have been built in Munich and Switzerland, and three more are under construction in Russia, Saudi Arabia, and San Diego, all being built by commercial companies.[146]

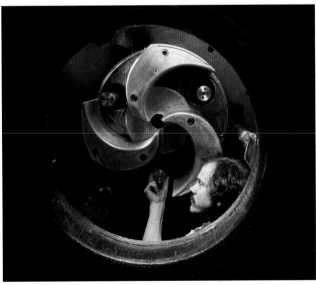

FIGURE 60. The Harper Cyclotron.

Opposite: The Cyclotron under construction in the S800 vault at the Cyclotron Laboratory. Blosser (*left*) and Powers are standing on the rotating rings of the gantry just above the cyclotron. The patient would lie on a table at the center of the rings. The neutron-producing target was inside the cyclotron, so there was no extraction problem. On the other hand, when the cyclotron was rotated, the reservoir holding the liquid helium that cooled the superconducting coil would be upside down. Building a reservoir that could be inverted and not spill required new thinking.

Top: The Cyclotron as installed at Harper Hospital in Detroit. The patient and physician need not be aware of the actual apparatus.

Bottom: Emanuel Blosser and the tiny magnet hills of the Harper cyclotron.

also behind schedule and the user community viewed it as a distraction from timely completion of the Phase II project. This led to some criticism of NSCL. Later, however, NSF took credit for the contributions of the project to treatment of cancer.

■ A New Way to Phase II: Don't Couple

A more important problem, however, was the Phase II cost overrun. It does not appear that its magnitude was fully appreciated until the NSF demanded an evaluation of the situation. The reasons for the cost overrun were detailed in a letter[147] from the NSCL to NSF. The cost-to-complete estimate was $43.9M, compared to a budget of $34.6M, and a spent-to-date amount of $24.2M. A major part of the overrun was due to the delays in the provision of federal funding, which increased costs, since the cost of maintaining manpower during delays does not disappear. But others had their origin with the project. Mainly, they were the results of needing to deal with unsatisfactory procurements; problems caused by the loss of hard-to-replace manpower on the radio frequency system due to deaths and other departures; and the delay and costs of obtaining a solution to the extraction problem.

In a site visit to MSU/NSCL in connection with the review of the March 1986 operating proposal, Harvey Willard, MSU/NSCL's NSF program officer, asked in a public meeting, "Why should NSF continue to fund a project that is so late and so over budget? Was success likely within an achievable budget?"[148] He certainly caught our attention. It is the author's opinion, based on conversations with colleagues outside the laboratory, that Willard's comments reflected a user-community concern with the delay of the Phase II project. And that this delay was partially due to competition for the laboratory's administrative and technical attention by the Harper Hospital neutron therapy project and other cancer therapy projects.

It was clear that the NSF did not want to cancel the project but was not willing to provide all the funds necessary for its completion. In the end, a joint solution to the funding problem was negotiated[149] that included contributions from both the NSF and MSU/NSCL.

This solution did not cover the entire funding shortfall and forced a decision to delay coupling the K500 and K1200 cyclotrons as had originally been proposed.[150] The beam line that was to couple the two cyclotrons (see figure 61) was long and complex, undergoing a U-turn after the exit from the

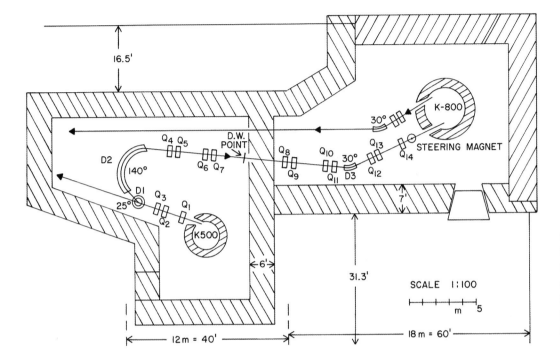

FIGURE 61. Original design for the coupling line leading from the K500 to the K1200. It is long and complicated, with several bending magnets and many focusing quadrupole magnets (denoted with Qs).

K500 before entering the K1200, and using a very large number of bending magnets and focusing quadrupoles. It would be an expensive device, and there was, as yet, no completely satisfactory design in hand for injection into the K1200, promising a further delay in an already delayed project.

The availability of the RT-ECR source offered a possible solution: begin operation with an RT-ECR + K1200 combination (no coupling) and construct a more powerful superconducting ECR. If it provided ions of sufficiently high charge, the design goals of the proposed coupled system could be met with a much simpler system. Figure 62 shows what would be achieved if the charge states from the Superconducting ECR (SC-ECR) were larger than those from the RT-ECR by various factors. Whether the SC-ECR would in fact produce such highly charged ions was far from certain, but a decision was made to take this approach.

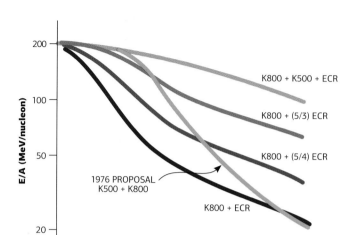

FIGURE 62. Performance of the ECR + K1200 system for various assumptions about the performance of the untested superconducting ECR (SC-ECR) source. For example, the curve labelled K800 + (5/3) ECR assumes that the charge on an ion of mass A produced by the SC-ECR is 5/3 times that from the RT-ECR and that the energy is therefore $(5/3)^2$ larger. The estimates of SC-ECR performance were educated guesses, but, if valid, the SC-ECR + K1200 system came close to matching the proposed 1976 specifications.

There were many advantages: earlier facility operation, lower cost, a simpler system, and a more reliable system. Given the low reliability of the K500, this latter advantage was, by itself, persuasive to the co-directors. A more complex coupled system would be less reliable than either of its parts, and even a perfectly reliable K1200 would yield poorer reliability than necessary to run an efficient program, unless the K500 was also significantly upgraded.

Another decision was whether to proceed directly to the final beam transport system or to use a small temporary experimental area (Phase I.5) located near the K1200 while the remaining beam lines were constructed.[151] The Phase I.5 choice might delay the final system by some months, but it had overwhelming advantages. Most importantly, it allowed early debugging and use of the K1200, a complex device that produced unique beams but might not work perfectly at first. If successful, it would also answer some of the nagging external questions about the project schedule. The beam line setup could also allow the K500 to operate and finish its PAC-approved experiments.

The possibility of providing secondary beams of radioactive elements had been in our minds for some time. It was first discussed at an NSCL Users Meeting held December 16–17, 1982 where Leigh Harwood presented the idea of using the beam transport line from the K1200 as a fragment separator to make radioactive beams. In 1984, calculations by Harwood, Brad Sherrill, and Jerry Nolen determined how well that would work and evaluated the use of absorbers to give additional selection capability. Sherrill presented these general ideas at a workshop at LBL in 1984,[152] helped design the GSI fragment separator while a postdoc at GSI in 1985–86, and then the A1200 fragment separator.

The A1200 was to be an S-shaped combination of dipole and quadrupole magnets that was located just after the K1200 cyclotron and would provide either beams with a well-defined momentum or beams of rare isotopes to following experimental stations. A design that repurposed magnets from the initially planned beam line reduced the construction cost to under $100,000.

The A1200 was an extremely cost-effective investment that initiated the use of radioactive beams at the NSCL. The A1200 and its successor, the A1900, strongly influenced the direction and focus of the NSCL nuclear science program for the next twenty-five years. Experiments with radioactive beams came to dominate the experimental program at the NSCL.

■ The Outcome

All of these decisions led to a facility different, but superior in many ways, to that originally proposed. It certainly made more efficient and reliable operation possible and opened up the study of radioactive beams. In this case, financial difficulty led to a positive outcome. The most unfortunate side effect was the necessity to postpone construction of the partially completed S800 superconducting spectrograph. It seemed certain that funding for this device would be obtained eventually, but that did not occur until 1993. The S800 was not completed until 1996.

The plan for the Phase I to Phase II transition then become:

- Phase I.5: Operate the K1200 cyclotron in a small experimental area containing the ninety-two-inch scattering chamber and the 4π Array (mid-October 1988 to February 1, 1990).
- During Phase I.5, construct the Phase II beam lines and new experimental areas, and operate the K500 cyclotron in the Phase I experimental area. Disassemble the K50 cyclotron (1989).
- After completion of Phase I.5, install Phase II beam lines, and move the ninety-two-inch

scattering chamber and the 4π Array to their final positions, freeing the space for installation of the A1200 fragment separator (February–April 1990).

- Begin operation of the K500 cyclotron with the Phase II beam lines (April 1990).
- Install the A1200 fragment separator (April 1990).
- Begin operation of the K1200 cyclotron with the Phase II beam lines and the A1200 fragment separator (October 1990).
- Discontinue operation of the K500 cyclotron except for accelerator physics tests (October 1990).

These decisions having been made, there remained the complex final assembly of the K1200 described in figures 63 through 67. This involved: multiple disassembles as previously described for the K500; completion of magnetic field measurements; and finally a solution to the radio frequency transformer power problem.

Internal beam was first observed at a radius of seven inches on February 13, 1988, and was accelerated to full radius on February 22. After a delay to install extraction hardware, a beam of $^{20}Ne^{3+}$ was extracted on June 6, 1988, with the calculated settings of the extraction elements.

After the years of angst, the extraction of the beam had proved to be simple and was announced in this rather low-key note, circulated by Blosser and Nolen on June 8.

FIGURE 63 (*opposite*). A rare view of the K1200 construction process. At bottom is the lower half of the magnet, showing the strongly spiraled hills; in the center is the copper "liner" that covers these hills and provides a vacuum in the space where the beam circulates; and at the top is the cryostat that contains the superconducting coils. The latter two parts fit into the magnet. Mirror images of the liner and magnet then complete the stack of elements that forms the magnet.

K800 EXTERNAL BEAM

On June 6, 1988, a beam of triply charged ^{20}Ne was extracted from the K800 (K1200) cyclotron at the National Science Foundation's National Superconducting Cyclotron Laboratory at Michigan State University. The beam came immediately through the extraction system with all of the 25 position adjustments set at precalculated values and with the strengths of the electric fields and of the first harmonic bump in the cyclotron very close to calculated values. The beam energy was 18 MeV/nucleon which means that the actual K value of the magnet was just at the original design goal of 800 MeV. (The maximum bending power available from the magnet now corresponds to a K of 1,200 MeV.) In the week prior to June 6th, beam had been detected at the end of the first electrostatic deflector, 60 degrees along the extraction path, but efforts to sense the beam in other locations along the total 330-degree path were unsuccessful. At the end of that week the cyclotron was opened to check the calibrations of the 25 position drives and one of the drives in the first electrostatic deflector was found

FIGURE 64. View of the cyclotron "high bay" during the construction of the K1200. The K500 is at the upper left and the K1200 at the lower right. The magnets themselves are at floor level, but the superstructures, mainly the radio frequency resonators that control the radio frequency, extend far above the magnets.

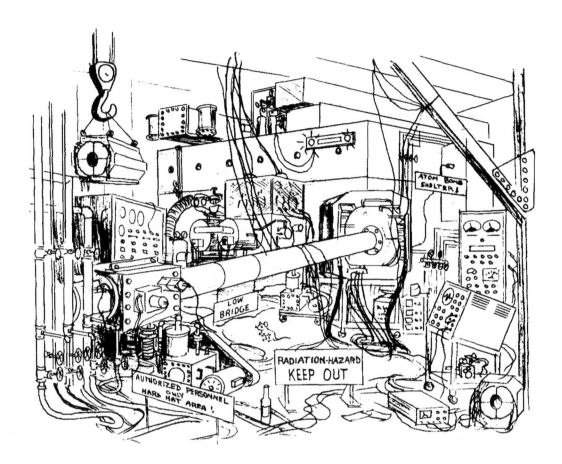

FIGURE 65. The apparent level of disarray, not real, seen in laboratories under construction is captured in this David Judd cartoon: "The Cyclotron as Seen by the Visitor."

to be miscalibrated by 4 mm. When operation resumed on June 6 with this error corrected beam was picked up successfully at the various check points through the extraction system and in an interval of no more than 20 minutes was on the exit flange of the cyclotron, with all current passing through a 1.5 cm square collimator hole. A later radio autograph of the end flange showed a circular beam spot approximately 4 mm in diameter. Operation of the cyclotron with other ions and at other energies will be explored in the coming weeks, in parallel with completing the beam line to the first experimental station. First experimental runs are expected in September.

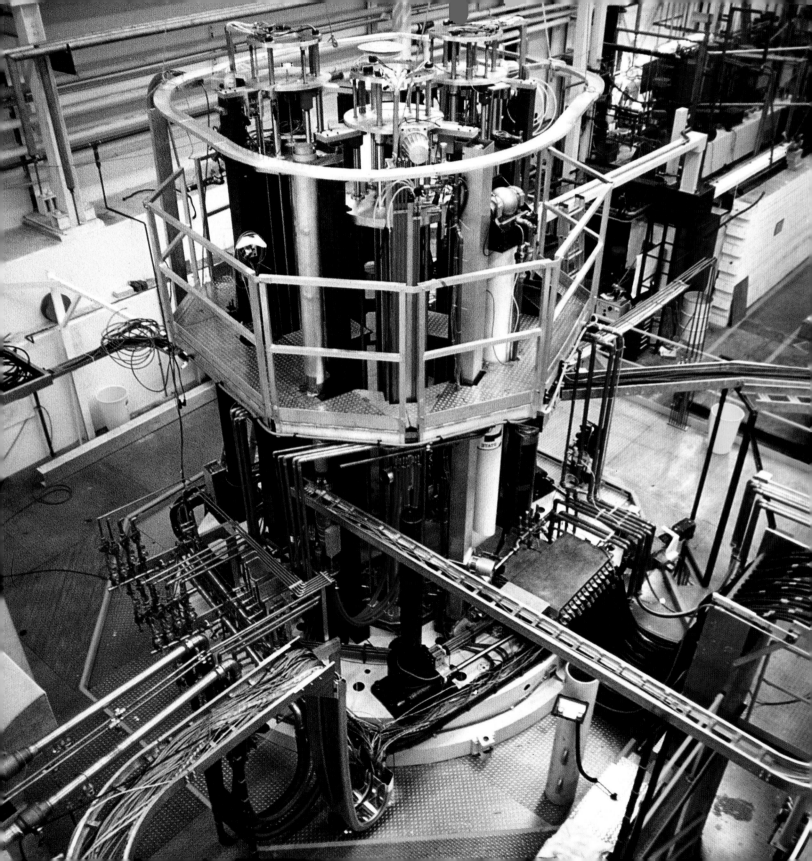

This and the ensuing development of other beams covering a significant part of the operating diagram gave confidence that beams interesting to experimenters could be provided. Henry Blosser, Felix Marti, Jerry Nolen, and John Vincent were all closely involved with beam extraction. Dave Johnson had calculated the extraction settings.

There were the usual teething problems associated with a new accelerator, and as had become usual, with the radio frequency system. By mid-1988, however, it appeared that a stable transformer design had been achieved and a spare transformer had been installed. A call for beam-time proposals went out in early July. PAC 9 met on September 18–20, 1988, and assigned time for experiments on both the K500 and the K1200 in its Phase I.5 mode.

The Phase I.5 vault housed a ninety-two-inch-diameter walk-in scattering chamber and the 4π Array. In mid-October 1988, the first experiment began, a measurement of high-energy gamma rays from collisions of 75 MeV/nucleon ^{14}N nuclei with Zn to search for effects of the Δ resonance of the nucleon. This was basically a shakedown experiment, to verify overall operational concepts. Production runs began on November 1.

It became necessary, however, to de-emphasize K500 operation in order to run Phase I.5 efficiently. There were four K1200 shutdowns to deal with radio frequency and cryogenics problems, but during the experiments themselves, breakdowns consumed less than 5 percent of scheduled time.

The removal of the K50 cyclotron during the Phase I.5 period (in 1989) was an emotional moment, especially for the first generation of researchers. The K50 had made the reputation of the laboratory and had made the NSCL possible. The magnet coils were later sent to Argonne National Laboratory to become part of the APEX experiment shown in figure 68. NSCL faculty and staff[153] played a major role in APEX.

When Phase I.5 ended in February 1990, twelve experiments had been completed, and it had been demonstrated that the K1200 produced high-quality beams. A total of thirty-three different beams (ion/energy combinations) had been developed, including ions from protons (in molecular combinations) to ^{129}Xe.

The K500 operated in January and August 1989 and then again during the K1200 shutdown for installation of the Phase II beam lines from February to September 1990. This time was used to complete K500 experiments previously approved by PAC 9.

FIGURE 66 (opposite). The K1200 near completion. This photo shows the contrast between the simple view of the cyclotron as seen in figure 4 and the working device. The magnet at the bottom is a large part of the weight of the cyclotron, but does not occupy most of the space. The radio frequency resonators (copper colored), the helium and nitrogen cryogen distribution pipes (white), and the copper cooling water pipes necessary to handle the heating by the radio frequency power make up most of the superstructure.

FIGURE 67. The K1200 and the people that built it; co-directors Blosser and Austin sit at the lower right.

FIGURE 68. The APEX experiment at ANL, a fourteen-institution collaboration with MSU playing a major role. The round thin coils were from the K50 and were used to form a uniform magnetic field. APEX measured spectra of electrons and positrons produced in collisions of very heavy ions, ^{238}U + ^{238}U for example. Experiments at GSI claimed to show that an electron and positron pair with a very well-defined energy was formed. This was a Nobel Prize–level "discovery," but, in the nature of the scientific method, APEX showed it was not true.

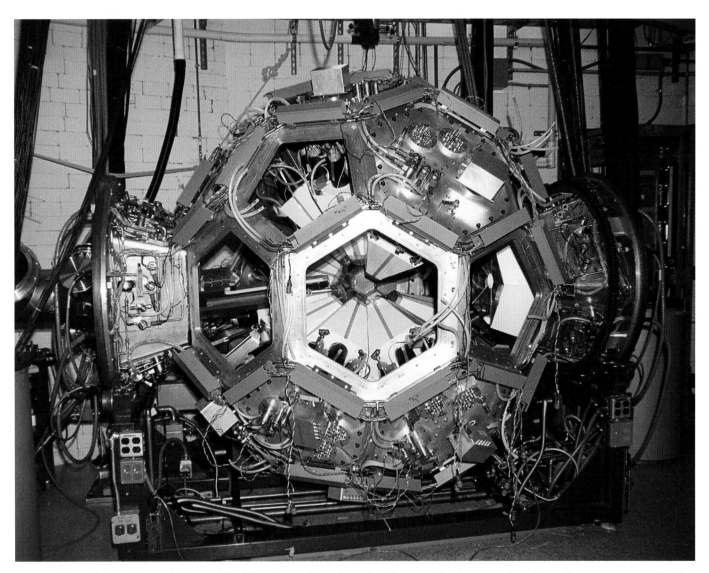

FIGURE 69. Views of the 4π Array: a device capable of detecting products of reactions that are emitted at any angle and range from protons to uranium. It has a soccer ball–like geometry that makes it simple to replace detectors at any angle with detectors with different characteristics. *Opposite*: The frame of the detector with its principal developer, Gary Westfall (*left*) and J. van der Plicht. *Above*: The array with its detectors, all looking at a target at the array's center. It could count particles at a very high rate.

FIGURE 70. Other complex arrays.

Top: A view of the Miniball, a portable detector that has been used internationally.

Center: HIRA, the High Resolution Array, with Prof. Betty Tsang (*right*), Sharon Suen, and other students are reflected in the gold surfaces of the silicon detectors.

Bottom: The Sega detector, with Dan-Cristian Dinca.

Opposite: Electronics for such detectors can be *complex*.

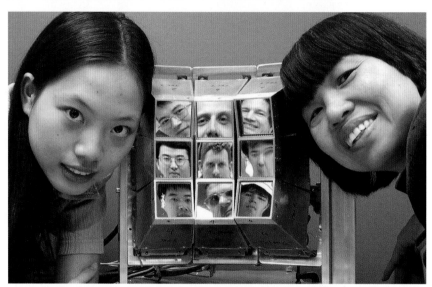

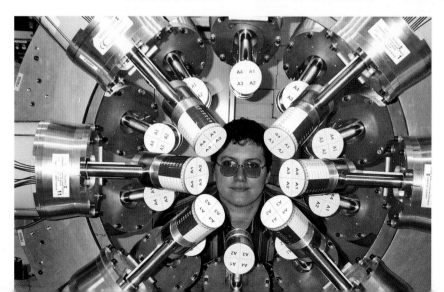

■ Problem-Solving, Transition, Reliability

During Phase I.5, construction of the Phase II beam line magnets, vacuum systems, diagnostic systems, new experimental areas, etc., was undertaken with the goal of completion by February 1, 1990. When the SC-ECR was completed in June 1990, three ECRs were available, giving time for development of new and difficult beams. New beam lines that made it possible to inject ions from any ECR (RT-ECR, CP-ECR, SC-ECR) into either cyclotron were completed.

Upon completion of Phase I.5, both cyclotrons were shut down, installation of beam lines in the transfer hall began, and the 4π Array (fig. 69); the ninety-two-inch chamber, which sometimes housed the Miniball (fig. 70); and the RPMS were moved to their final experimental areas. In April 1990, operation of the K500 cyclotron began again while the A1200 fragment separator was installed. This was completed in October 1990, and operation of the K1200 began. A measurement with the A1200 showed that the energy spread of the K1200 beam was 0.05 percent, exceptionally good for any cyclotron.

A solution to the nagging problem with the K1200's radio frequency system, involving the decreasing quality of RCA final amplifier tubes provided by the manufacturer, also began during the shutdown.[154] After these changes were completed in 1991 and 1992 and the K1200's central region was redesigned to reduce voltage requirements, radio frequency problems were, mainly, a thing of the past. The reduced voltage also reduced power consumption by 40 percent, a major cost saving.

By then, PAC-approved experiments on the K500 were essentially complete, and the K500 was placed on cold standby (i.e., the coils were kept at liquid helium temperature) until its future was decided. Eventually, the K500 would play an important role in the next phase of NSCL's evolution, but, although scientific arguments for continuing operation of the K500 were strong,[155] the NSF decided not to provide the necessary funding. PAC 9 was the last to consider proposals for solo K500 time. It was used on occasion to validate accelerator concepts, until September 1996 when it was shut down for disassembly and refurbishment to prepare for the new Coupled Cyclotron Facility (CCF).

At this point the main issue was to increase reliability of operation with a goal of reducing breakdown time to less than 10 percent of scheduled time. A Committee on K1200 Operations (COKO) was formed in November 1990 and was charged to determine whether this was a realistic goal and what changes would help improve efficiency.

Suggestions for increasing beam energy and intensity were also sought. A detailed report

FIGURE 71 (opposite). The S800 spectrograph in two views:

Top: The conceptual design. The intermediate stage (square boxes are bending magnets and the other elements are focusing quadrupoles) can be used in several ways to prepare the beam for later detection in the S800. The detectors are located in the focal plane.

Bottom: As actually used, in this case with a large gamma-ray detector called GRETINA, built mainly at the Lawrence Berkeley National Laboratory. The S800 fills a three-story experimental vault. It is the most widely used instrument in the laboratory.

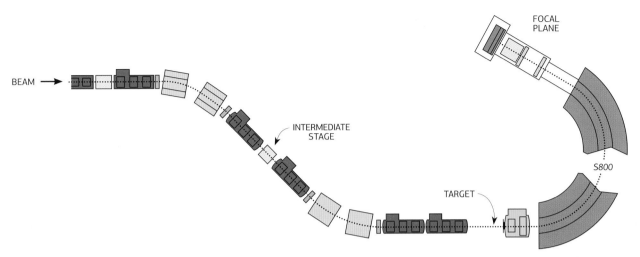

was provided in April 1991, and by June 1992, a number of the recommended changes had been implemented: those changes affecting users were scheduling three months in advance of their use with a requirement to submit a description of the experiment needs prior to consideration for scheduling. A program for providing backups for equipment whose failures would produce long downtimes was in place and procurements were being made as funds became available. Because of these steps, and because the early start-up failures were in the past, reliability soon increased to the 90 percent level required to run an efficient user program.

One major piece of equipment was still in the pipeline: the partly finished S800 spectrograph shown in figure 71. This large device bends the products of a reaction in a vertical direction. Its combination of resolution and ability to accept reaction products emitted in a large range of angles (large angular acceptance)[156] is still unmatched. It was also unusual for its time in that it did not owe its excellent resolution to elements that corrected for aberrations, optical effects that decreased the resolution; this had proved impossible for the large angular acceptance of the device. Instead, detectors at the end (focal plane) of the S800 measured the position and angle at which each particle arrived at these detectors. Using the precisely measured magnetic field of the S800, the particle trajectories could be traced[157] back to the target, yielding the angles and energies at that point. Because of the large angular acceptance, most reaction products of interest are detected. Consequently, the spectrograph has almost never (only once) been rotated away from zero degrees.

The S800 had been part of the original Phase II proposal, and some major items, for example, the steel for the large dipole magnets, had been purchased. During Phase II construction, however, S800 completion was postponed and ECR sources and a beam analysis device, the A1200, were built instead. In the interim, the NSF had found it difficult to obtain funds for S800 completion.

MSU organized an International Conference on Heavy-Ion Research with Magnetic Spectrographs,[158] held in January 1989, and in July submitted a new proposal to NSF.[159] Funding to complete one dipole was obtained soon thereafter, and final NSF funding to complete the S800 arrived in 1993. The S800 was commissioned in September 1996 and has become the most used instrument in the laboratory, both in its high-resolution mode and as an efficient collector and identifier of reaction products produced by rare isotope beams.[160]

The Next Step: Coupled Cyclotrons Again

In July 1986, during the review of the 1986 operating proposal, Sam Austin, then NSCL co-director, stated that at the completion of the K1200 cyclotron, planning for the long-term future of the NSCL needed to begin. It took longer than was then anticipated to finish the K1200, but even before the K1200 + ECR system was running smoothly, planning for possible new accelerator systems at the NSCL had begun, and by 1992 was in full swing. At first glance, such early planning seems unusual, but it was driven by a combination of opportunity, time scales, and competition.

■ The Situation: Why We Had to Act Quickly

The priority recommendations of the 1989 NSAC Long Range Plan (LRP-1989) were mainly for high-energy facilities: CEBAF, an electron accelerator facility (now JLAB) in Newport News, Virginia; RHIC, a colliding beams facility for heavy ions at Brookhaven National Laboratory; and possible participation in a high-intensity proton accelerator (KAON) at TRIUMF in Canada.

Only the plan's fourth recommendation concerned lower energy facilities. In rather generic terms, it listed some of the proposals that were being considered for the future and "anticipated that at least one such project will achieve high scientific viability over the period of this LRP."

Later in the report, some of these possibilities were described in more detail. First on the list was a radioactive beam accelerator. The plan did not state that this was in order of priority, but as is common in such documents, this was implied/assumed.

The LRP also enumerated a five-point "agenda" for nuclear science. Two of its points—"Study of the thermodynamic properties of nuclear matter, expressed in the equation of state, and its phase transitions" and "Searches for new phenomena at the very limits of nuclear stability"—were precisely those that could be initiated with the K1200 + ECR facility and much better met by an improved radioactive beam facility. Not specifically covered in LRP-1989, but discussed in the Town Meetings leading up to it, was the concept of a major facility for the (far?) future, possibly after RHIC was finished.

Certainly, LRP-1989 provided an opening for a near-future NSCL upgrade. But it also set a time scale: if an MSU radioactive beam facility was not proposed before the next LRP, another facility probably would be and, if judged attractive, would probably be first in the queue for funding. The next long-range plan was due in about 1995, so if a proposal wasn't ready for consideration by then, it would be 2000 before possible NSAC approval, then at least two years to appear in the federal budget, and if approved for funding, at least another five years to build the accelerator—if all went well. That meant it would be 2007, at least, before a new facility would operate, and this could easily slip to 2010 or later. The LRP considerations also set a cost scale for a near-term proposal: it could not be a major (i.e., hundreds of millions of dollars) facility. Given the other priorities, funds were just not available.

There was also the issue of competition. Several radioactive beam facilities, at GANIL in France, at GSI in Germany, and at RIKEN in Japan, would soon provide beams superior to the NSCL in intensity and/or energy. On the U.S. scene, it was becoming likely that NSF funding for nuclear physics would soon be unable to support both the NSCL and the Indiana University Cyclotron Facility (IUCF). Since IUCF would be the likely proponent of one of the other facilities on the NSAC LRP list, it appeared that whichever of the two labs was successful in this 1995 round of proposals would be more likely to have a long-term funding future.

That this was the case became clearer in January 1992, when NSAC formed a subcommittee[161] chaired by John Schiffer of ANL, in response to an NSF/DOE charge to provide guidance on nuclear science priorities under constrained budgets. The subcommittee was to consider three options: 2 to 3 percent growth beyond inflation, constant effort or just inflationary increases, and constant dollars or a decrease in effort by the amount of inflation. Most thought the last option was the most likely.

The subcommittee heard oral presentations of laboratory achievements and future directions by both laboratories. The reviews of both the NSCL and IUCF programs were strongly positive, but the committee recommendations caused concern. They were: under the growth scenario, constant effort for NSCL and IUCF, allowing the agencies to undo earlier cuts in support for users and smaller facilities; under the constant effort scenario, reductions in one of these facilities; under the constant dollar scenario, a review—the context of the document implying that either MSU/NSCL or IUCF would have to be closed.

Shortly after this review, Konrad Gelbke was appointed Director of the NSCL, replacing Sam Austin who had chosen not to seek another term as director. Gelbke had been at MSU/NSCL since 1977 and had compiled an enviable record of research in studying the collisions of heavy ions. He began as Director at a challenging time for the NSCL but as yet had little experience in that role.

One of his first tasks was to participate in yet another NSF review and carry out its recommendations. As budgets tightened, in late 1992 the NSF asked NSAC for specific guidance on how to apportion 5 percent and 10 percent cuts in the total funding for NSCL and IUCF. Again, NSAC formed a subcommittee, this time chaired by Robert Redwine of MIT, that performed a detailed review of both laboratories. The subcommittee in February 1993[162] recommended that:

FIGURE 72. Konrad Gelbke at the time of his appointment as NSCL Director.

> Given the excellence we found at each laboratory, we do not find justification to recommend that one be cut significantly more than the other. We are also guided by the need to maintain flexibility to recover in the event future years allow more positive funding situations. We therefore recommend that NSF apportion necessary budget reductions at IUCF and NSCL in FY93 roughly equally, consistent with the priorities listed above.

As a result, in mid-1993, NSCL funding was cut by about 12.5 percent, forcing large reductions in personnel and services to users.

■ What to Build?

Given all these considerations, the characteristics of a new facility were constrained. It needed to produce intense radioactive beams and also beams sufficiently heavy and of sufficient energy to facilitate studies of the nuclear equation of state.

It was less clear what the new facility should be. Three options were considered in enough detail so that approximate costs and performance could be estimated to answer the questions: does it do what it needs to do *and* is it rational to expect funding in the present financial situation?

The first two options—building a new rapid-cycling synchrotron injected by the K1200, or building a large new separated sector K2000 cyclotron injected by the K1200—did not appear to meet these criteria.

The K1200 + rapid-cycling synchrotron option, developed by Richard York, would be relatively inexpensive and could produce beams in the (one to two) GeV/nucleon range, an advantage for the equation of state applications but of too-low intensity for radioactive beams. Nor would it compete adequately with the large synchrotron being built at GSI as part of its FAIR project.

The K2000 + K1200 option, examined by Felix Marti, would provide the desired beams, with energies in the 400 MeV/nucleon range. It would have a separated sector form,[163] and the facility cost would be in the $100M range. This was the initial MSU choice, but during a visit to NSF to discuss this possibility, MSU was told that NSF would not fund such an expensive facility.[164] NSF outlined criteria that a future proposal would have to include: total cost to NSF of around $10M, an MSU cost share, and a redirection of some funds from the operating budget. The NSF representatives also stated that they were not ready to consider a formal proposal.

At that point, the third option considered, ECR + A1200 + K1200, emerged as a viable alternative. It was based on a concept proposed by Felix Marti[165] and resembled the coupled cyclotrons of the 1976 proposal but with a decisive advantage: injection to the K500 by an ECR source. An intense beam of low-charge-state ions from the ECR would be injected into the K500, accelerated, extracted from the K500, and injected into the K1200. The net result of this procedure was an increase of primary beam intensities by factors of ten or more (depending on energy) compared to the ECR + K1200. Higher energies, almost 100 MeV/nucleon for ^{238}U, were also possible. Using these beams with a more powerful version of the A1200 fragment separator, called the A1900, would yield gains of 1,000 to 10,000 in rare isotope beam intensity, enough for a great expansion of research with radioactive beams. This satisfied an informal but important criterion: a new accelerator has to promise greater than a factor of ten improvement in the most crucial criterion.

To achieve these gains required a number of changes to the K500 and K1200, and the studies necessary to delineate them took some time. In July 1994, a White Paper, MSUCL-939, "The K500 + K1200, A Coupled Cyclotron Facility at the National Superconducting Cyclotron Laboratory, Michigan State University," was widely distributed for comment. In spite of NSF's statement that they were not ready to consider a formal proposal, NSCL and MSU decided to force the issue, and in September 1994, a proposal, MSUCL-949 requesting $11.0M, was submitted to NSF to partially fund the construction project. MSU would provide both the high bay necessary to house the project assembly area as a start-up cost and a $7M cost share, *provided* the NSCL operating budget from NSF was increased to cover inflation. In addition, $3M from the operating budget of NSCL was used (redirected) to support construction of the CCF.

This proposal received strong positive reviews from NSF reviewers, but NSF decided it would wait for the recommendations of the next NSAC Long Range Plan before considering whether to provide funding.

■ The NSAC Process and the Coupled Cyclotrons

In September 1994, NSAC was charged by DOE and NSF to provide advice for the future evolution of nuclear science. Six Town Meetings organized by the American Physical Society's Division of Nuclear Physics (APS/DNP) and a number of White Papers on different topics, including the NSCL Coupled Cyclotron White Paper, provided input for the meeting of the NSAC Long Range Plan (LRP) working group in Pasadena, March 16–21, 1995. The resulting recommendations were published in February 1996 as the "Nuclear Science Long Range Plan" (LRP-96), but the recommendations were known informally much earlier, and were (paraphrased, except for numbers three and four):

1. Provision of funds to use the existing infrastructure efficiently.
2. Completion of The Relativistic Heavy Ion Collider (RHIC) at Brookhaven National Laboratory as the highest construction priority.

3. "The scientific opportunities made available by world-class radioactive beams are extremely compelling and merit very high priority. The U.S. is well-positioned for a leadership role in this important area; accordingly:

- We strongly recommend the immediate upgrade of the MSU facility to provide intense beams of radioactive nuclei via fragmentation.
- We strongly recommend a development of a cost-effective plan for a next generation ISOL-type facility and its construction when RHIC construction is substantially complete."

4. "We strongly recommend funding for a Light-Ion Spin Synchrotron (LISS) as a major NSF research initiative . . ." and "collisions of polarized proton beams in RHIC . . . should be pursued."

Even though LISS backers had not put together a detailed proposal, there was a last-minute move at the LRP working group meeting to move the priority of LISS above that of the MSU Coupled Cyclotron Facility. It was motivated by the recognition by some attendees that the future of the Indiana University Cyclotron Facility was probably at stake.

Their move was not successful, and funding for LISS construction became unlikely. As LISS proponents had feared, this led, eventually, to closure in 2003[166] of the IUCF accelerator and its cooler facility for research in nuclear physics, and in 2014 termination of a cancer therapy program using its cyclotron that had been established at IUCF.

Both parts of the working group's "recommendation" had a strong effect on the future of the NSCL. Well before the formal publication of LRP-96, NSF had formed a high level technical review panel to assess the MSU Coupled Cyclotron proposal. In its report of July 1995, the panel concluded:

> The quality and thoroughness of the work presented was impressive, leaving the panel convinced of the general feasibility of the project and of the credibility of the proposed cost estimates and schedule. No unreasonable technical risks were identified. NSCL staff seem to be aware of all the major technical problems and to be taking the appropriate steps to solve them. With care, the project specifications should be achievable, opening up a rich vein of new nuclear physics in an energy and intensity regime not accessible by other facilities.

With this blessing, NSF approved funding for the Coupled Cyclotron Facility in November 1996, and first funds arrived in the first quarter of 1997. Some work, including construction of the high bay, began earlier, using funds from the MSU cost share. This significantly advanced completion of the project.

■ Building and Operating the Coupled Cyclotrons

There were several major steps in the CCF project as shown in figure 73. The K500 cyclotron had to be refurbished and made more reliable. It had to be rotated to provide a simpler coupling line to the K1200 cyclotron. The many details of coupling had to be designed, constructed, and installed. And a new ECR source had to be built. It was expected that the process would proceed smoothly because solutions to known problems were in hand.

The K500 was first run to validate concepts for the CCF, for example, a new second harmonic central region, a new spiral inflector for injection, and testing of beam phase diagnostics. By September 1996, testing was complete, and the K500 was shut down for disassembly and refurbishment. This involved many changes to improve capability and reliability, in most cases implementing concepts proven on the K1200. By July 1997, these changes were complete, and the K500 had been reinstalled in its vault and rotated by 120 degrees in order to send its beams to the K1200 rather than directly to experiments. Compared to the original Phase II design, the rotation greatly simplified the injection into the K1200. Magnetic field mapping, new radio frequency amplifiers, and extraction elements remained to be completed. K500 commissioning was finished on December 1, 1998.

New and more powerful ECR sources were also built: a room temperature source, ARTEMIS A, that would be the main operating source, and later a new superconducting source, SuSI.

Building the new A1900 fragment separator, figure 74, was a major project led by Dave Morrissey.[167] For scheduling and cost reasons, one strong recommendation of the Review Panel was not implemented: use of so-called cos2θ quadrupoles that would have increased the efficiency of the device, rather than use of NSCL-standard iron-dominated quadrupoles. Nevertheless, the A1900 has turned out to be an extremely powerful and successful instrument.

Because the new A1900 fed all beam lines leading from the coupled cyclotrons, a shutdown of experimental activity lasting eighteen months was needed to remove the A1200 separator and install the massive A1900 separator. This would be followed by a six-month commissioning phase. The shutdown began as scheduled on July 1, 1999.

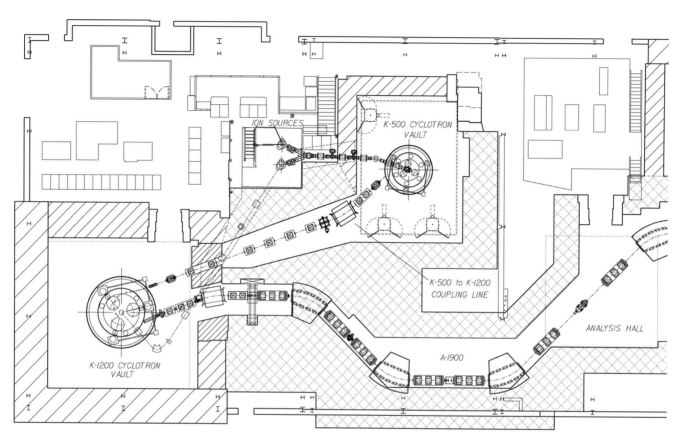

FIGURE 73. *This page:* The coupled cyclotron facility (CCF). *Opposite page:* The Phase II, K1200 + ECR + A1200 facility, is shown for comparison. The ECR sources and their switchyard are just to the left of the K500. The changes are evident. During Phase II, the K500 and K1200 operated independently. In the CCF the K500 has been rotated and now injects its beams into the K1200 through a short straight injection line—compare with figure 61. Since CCF beams are much more intense than Phase II beams, much more massive radiation shielding is required. The A1900 is much larger than the A1200, it accepts a larger fraction of the rare isotopes produced, and in general has much greater capability.

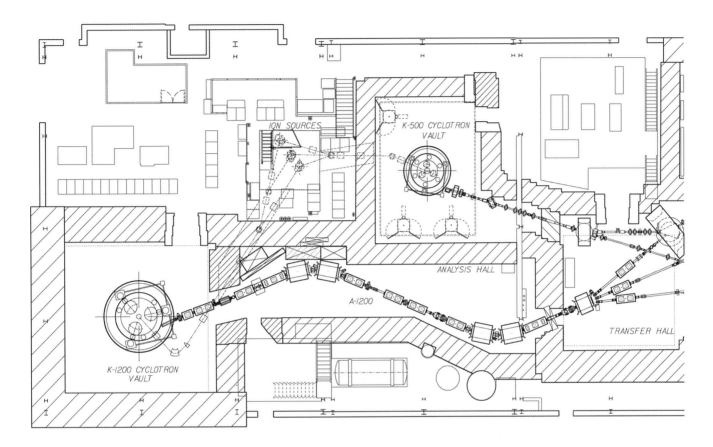

From there, developments came rapidly.

- October 2000: The first beam from the coupled cyclotrons was extracted.
- April 2001: Commissioning of the A1900 began.
- May 2001: The first rare isotope beams were observed.
- June 2001: The first experiment was performed.

The Coupled Cyclotron Facility (CCF) project was carried out on time and essentially on budget, by a team led by Richard York.[168] With the CCF, formal project planning techniques were used

for the first time at the NSCL. These are now an essential part of the FRIB project planning and management process.

As with any new accelerator system, a number of problems and bugs appeared during early CCF operation. These are more serious when one has two cyclotrons: if each operates with 90 percent reliability, the pair will have 81 percent reliability, assuming no common mode failures, so one must pay closer attention to phenomena that cause breakdowns. A reliability improvement system was instituted that involved rigorous failure reporting and follow-up with failure-cause determination, and major investments were made in replacing old failure-prone equipment, especially power supplies. This system was put in place by then-Associate Director for Operations, Thomas Glasmacher. NSF funding was also obtained for a major cryogenic upgrade carried out with the assistance of JLAB.

FIGURE 74 (*opposite*). Building the A1900. The green cylinder near operator Kevin Edwards contains three focusing quadrupole magnets, There are two of these triplets near the letters A1900 in figure 73.

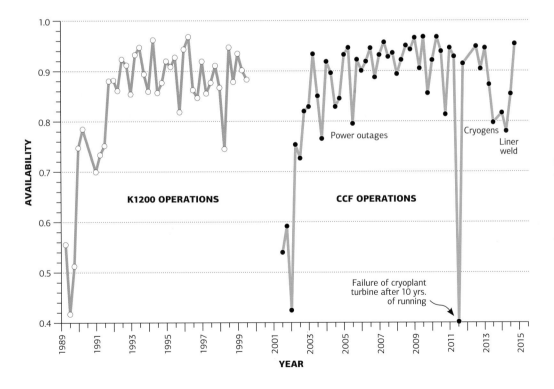

FIGURE 75. Reliability statistics for the K1200 Cyclotron during Phase II and for the CCF. After a startup phase, the accelerators usually operated with greater than 90 percent reliability (0.9 on the graph), as is required for an efficient user program. Occasional unanticipated events do, however, temporarily reduce reliability.

FIGURE 76. *Left*: Undergraduates assembling MoNA. *Right*: Prof. Artemis Spyrou at the controls.

By 2004, when chronic radio frequency system failures were finally dealt with, reliability reached 90 percent, and it has remained at that level with occasional short-term exceptions. The K1200 has now been operating for over twenty-five years, and some signs of old age—fragile welds, for example—have appeared, and led to short-terms problems (see figure 75), but they have been repaired and do not, at present, significantly limit K1200 operation.

With adequate reliability achieved, more attention has been devoted to improving the efficiency of operating procedures. As a result, substantially less time is required to produce a given beam, beams from the ECR sources are more intense, and the efficiency of beam transport through the two cyclotrons has increased.

The latter process was greatly enhanced when a new ARTEMIS B source began offline operation. This source, an exact copy of the main operations ARTEMIS A source, made it possible to study source and beam line behavior much more effectively than with an online source, which always has the priority task of providing beams for experiments.

In 2009, the first superconducting ECR, SC-ECR, was replaced by SuSI, a superconducting ECR that is able to operate at higher radio frequencies and achieve greater beam intensity. Eventually a combined coupling and extraction efficiency of about 90 percent was obtained for a ^{48}Ca beam.

A number of changes were also made in the suite of experimental instruments and the layout of experimental areas to accommodate them. After its last scheduled experiment, the 4π Array detector was decommissioned in December 2006, and is now on display in the atrium of the Biological-Physical Science Building. Its vault is now devoted to preparation of stopped beams for mass measurements and reacceleration in the superconducting linear accelerator ReA3.

The RPMS spectrometer was now obsolete, and the Rochester Superball (see "Detection Apparatus Generations") had performed its highest priority experiments by 2005. These devices were removed to make room for a radio frequency fragment separator, which sorts isotopes by their velocity, resulting in much lower backgrounds for experiments with isotopes having large Z/N ratios; a new fifty-three-inch-diameter scattering chamber; liquid scintillator–based neutron detectors;[169] a large scintillator array (MoNA-LISA)[170] for neutron studies (constructed and used by a collaboration with many undergraduate institutions); and the LEBIT system[171] built by Georg Bollen for mass measurements. LEBIT uses a gas-filled device to stop fast ions followed by a precision Penning trap and was able to measure nuclear masses with a precision of one part in 100 million.

Many other smaller pieces of experimental apparatus have been developed to optimize experiments with radioactive beams from the coupled cyclotrons and the A1900. These are listed in the "Present Coupled Cyclotron Era" part of "Detection Apparatus Generations."

FIGURE 77. Thomas Glasmacher earned MS (1990) and PhD (1992) degrees in low-energy experimental nuclear physics from Florida State University and came to NSCL as an NSCL Fellow in 1992. In 1995, he joined the MSU faculty in the Department of Physics and Astronomy and NSCL, where he is now a University Distinguished Professor. His research focused on exploring the structure of rare isotopes with new experimental techniques involving gamma-rays; he built and used the gamma-ray array SEGA. This work was recognized in 2006 with the Sackler Prize in the Physical Sciences. He is now Project Manager and Director for the FRIB Project and Director of the FRIB Laboratory.

The Next Big Thing

In the 1990s, nuclear physicists worldwide became convinced that the most promising approach for gaining an understanding of the fundamental nature of the atomic nucleus lay in studies using rare isotopes. This required intensities and a variety of isotopes that were well beyond those produced at the NSCL Coupled Cyclotron Facility and would require construction budgets well beyond its costs. The principal issue was choosing between the two most promising techniques—projectile fragmentation and the Isotope Separation On Line (ISOL) method—for producing these beams.

■ Projectile Fragmentation Versus ISOL

In the projectile fragmentation (PF) method, a high-energy beam of heavy nuclei bombards a thin target made of light nuclei, and the many product isotopes from the nuclear reactions go forward with roughly the same direction and speed as the beam striking the target. There may be hundreds of such products, so a fragment separator like the A1200 or A1900 is needed to sort out the nuclei of interest. These fast nuclei are then used directly as beams for experiments.

In the ISOL method, a high-energy beam, usually protons, stops in a thick target composed of heavy nuclei, the many isotopes produced diffuse out of the target, and are then sorted, collected, and accelerated for use as beams.

PF method beams are produced with high energy but are less precise than ISOL beams. However, they are better than ISOL beams for studying nuclei with very large or very small ratios of protons to neutrons, or short lifetimes. For various technical reasons[172] they provide a far greater scientific reach toward rare isotopes with extreme ratios of neutrons to protons.

ISOL method beams can be precise, are a good match for the present generation of low-energy accelerators, and have higher intensity for certain isotopes. They also have disadvantages: the radioactive nuclei to be accelerated are produced at rest and may be difficult to extract rapidly and efficiently from the beam stopper because of unfavorable chemical properties. A large and expensive accelerator would be required to accelerate them to high energy, so proposed ISOL facilities have not chosen to provide a high-energy capability.

MSU-NSCL championed a new paradigm, a combination of these techniques that proved uniquely powerful: the fast PF beams can be used directly or can be stopped and then reaccelerated in a superconducting linear accelerator.

■ The Evolution of RIA and FRIB

An early sign of U.S. interest in the science of rare isotopes was the formation of the IsoSpin Laboratory (ISL) Steering Committee and the following 1991 ISL White Paper: "Research Opportunities with Radioactive Nuclear Beams."[173] This White Paper recommended construction of an ISOL facility for studies with relatively low-beam energies of up to 10 MeV/nucleon.

In the following years, the international community of nuclear scientists, through a series of White Papers and meeting proceedings, played a large role in making the case that RIA/FRIB could address crucial issues in nuclear science.[174] Many of these documents produced over a span of twenty years have been summarized by the FRIB Users Organization and are listed on their website: http://fribusers.org/2_INFO/7_history.html.

The NSAC Long Range Plan (LRP-1996), mentioned earlier, also supported a next-generation ISOL facility, effectively assuming that research with fast beams would, for the foreseeable future, be adequately performed at the NSCL Coupled Cyclotron Facility or overseas. This recommendation

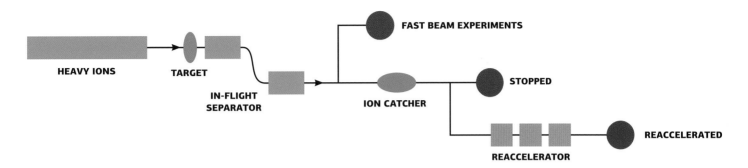

HEAVY IONS

TARGET

IN-FLIGHT
SEPARATOR

FAST BEAM EXPERIMENTS

ION CATCHER

STOPPED

REACCELERATOR

REACCELERATED

was reiterated in the 1999 National Academy of Sciences Decadal Report, *Nuclear Physics: The Core of Matter, The Fuel of Stars.*

In 1998, NSAC charged a task force including several from NSCL[175] and headed by Hermann Grunder, to: "provide a technical analysis of the various options for subsystems of a new facility for a research program along the lines indicated by the benchmark experiments outlined in the 1997 physics report, 'Scientific Opportunities with an Advanced ISOL Facility.'"[176]

Following a suggestion of Konrad Gelbke[177] to the committee, MSU was charged with evaluating the option of slowing fast beams and reaccelerating them. During a task force meeting in March 1999, Brad Sherrill of the NSCL gave the results of this evaluation in a presentation on the properties of rare-isotope beams that could be produced by the PF method from primary 400-kilowatt uranium beams. He pointed out that after slowing the fast beams and reaccelerating them, the intensities were comparable to those obtained by the ISOL method without the limitations described above.[178] And, as a bonus, the fast PF products would be available for experiments. Figure 78 summarizes this approach.

These conclusions were mirrored in the ISOL Task Force Report to NSAC on November 22, 1999, and the federal agencies on December 3, 1999. The Report recommended a facility based on a superconducting heavy-ion linear accelerator that combined the characteristics and advantages of the PF and ISOL techniques as described above. The projected cost of $500M for this Rare Isotope Accelerator (RIA) would turn out to be an underestimate for an entire facility.

NSCL task force members had produced a White Paper arguing for use of a large cyclotron to produce the primary beam. When the Task Force Report recommended a facility based on a

FIGURE 78. Diagram of the proposed RIA process. Heavy ions from the superconducting linac are incident on a target and the many fast-moving isotopes produced are sorted by a fragment separator, an enlarged version of the A1900. They may then be used directly for experiments, or captured and stopped in an ion catcher for use in experiments with isotopes at rest. The stopped ions may also be reaccelerated and used for experiments requiring precise low-energy beams. This approach combines the characteristics and advantages of the PF and ISOL techniques.

superconducting heavy-ion linear accelerator for producing the primary heavy-ion beam, an MSU group led by Richard York immediately undertook the design of a linear accelerator system, and design and prototype testing of the superconducting radio frequency cavities that it would require.

■ What's Next for the NSCL?

NSCL was the world's most productive source of research with rare isotopes. Its research into their nuclear properties and the role they played in the evolution of stars led the world. But it appeared that its long-term survival as a world-leading laboratory was again in doubt. Nuclear science is highly competitive, and accelerators were being developed in Japan and Europe that would soon challenge the NSCL's preeminence in the field.

It was far from clear how MSU would meet this challenge. Competing for the next major nuclear science facility, an accelerator system on the billion-dollar scale, would call for a great change in the nature and structure of MSU-NSCL. And MSU would face strong competition from National Laboratories for the right to build this facility.

Following Sherrill's presentation at the ISOL Task Force in March 1999, opinion in the NSCL crystallized around a decision to compete for RIA while completing the CCF. MSU's main competitor would be Argonne National Laboratory. The true strength of the MSU effort was hard to assess. Based on its research record with radioactive beams from the A1200 fragment separator and the upcoming coupled facility, MSU had the better case. Expertise and accomplishments in nuclear astrophysics, as reflected by NSCL's large role in the Joint Institute for Nuclear Astrophysics (JINA), were also a plus for MSU.

Soon thereafter, on December 8, 1999, an MSU group including Konrad Gelbke, MSU President Peter McPherson, and other MSU representatives gave a presentation to DOE and NSF about "The Benefits of Siting RIA on the Michigan State University Campus."

But many at DOE, the future funder of RIA, appeared to feel strongly that this facility should be at one of its established national laboratories, partly because of doubts that a university lab, even one as large as MSU-NSCL, could succeed in building such a large project. Moreover, ANL had expertise in superconducting linear accelerators that, at the time, did not exist at MSU. The Argonne staff felt that their expertise made them the strongest candidate for this facility. There

FIGURE 79 (*opposite*). One of the first MSU investments was for a large clean room to provide a space with a very small amount of particulate matter in the air. This is crucial for the production of the niobium radio frequency cavities that provide the acceleration voltages for FRIB, as well as for other accelerator components. Here work is on a section of beam line.

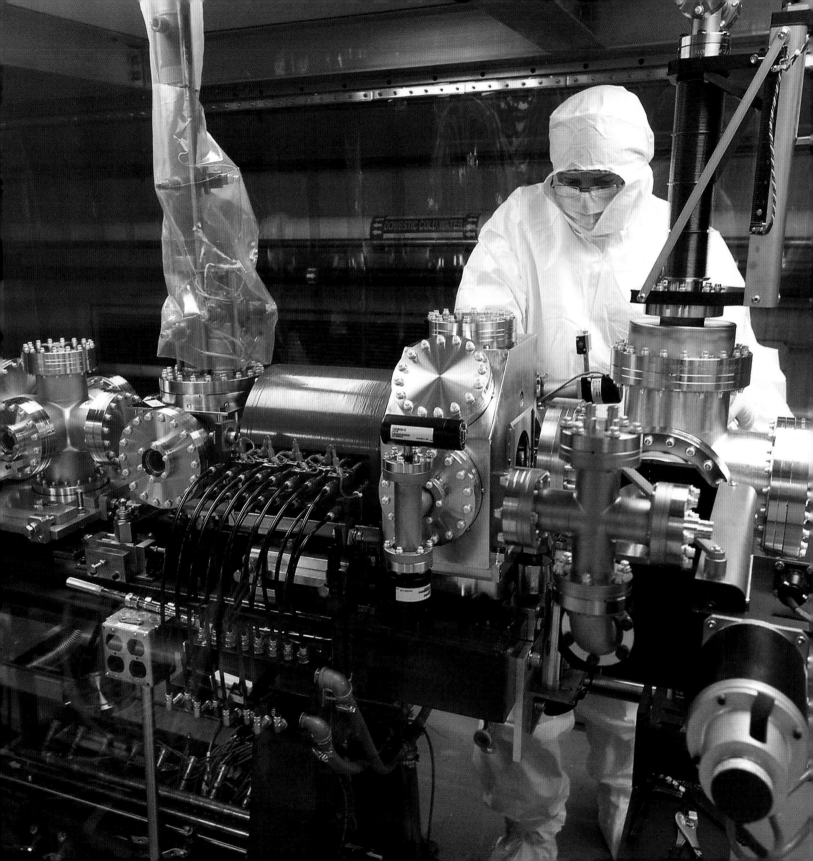

FIGURE 80. Preparing stopped beams.

Here: The rare isotope beam from the CCF or FRIB enters from the lower right. It first passes through a solid absorber that reduces its energy but produces a large energy spread. A magnet system then reduces, or compresses, this spread and leads the beam to one of two stoppers, where it is stopped in helium gas and then extracted for injection into ReA3.

Opposite: The cyclotron gas stopper uses a new technique invented and developed at the NSCL. It might be called an inverse cyclotron. The beam enters at large radius, loses energy in the helium gas that fills the cyclotron, and spirals toward the center, where it is collected for injection into ReA3 or use in other experiments. The stopper is a large device; Chris Magsig, magnet engineering team leader, provides a scale. It does not have, or need, a radio frequency system.

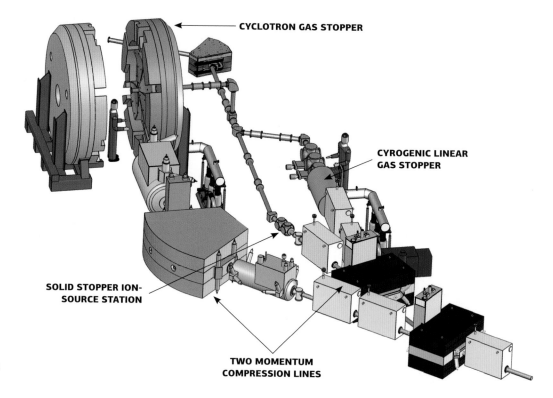

CYCLOTRON GAS STOPPER

CYROGENIC LINEAR GAS STOPPER

SOLID STOPPER ION-SOURCE STATION

TWO MOMENTUM COMPRESSION LINES

were intense, often quite negative and unproductive interactions of ANL and MSU researchers that peaked at a meeting of the Division of Nuclear Physics in Asilomar, California, in fall 1999, and continued for some time.

In this situation MSU had to establish three facts:

- That the physics with fast PF beams was of crucial importance. A White Paper, "Scientific Opportunities with Fast Fragmentation Beams from the Rare Isotope Accelerator," issued by MSU in March 2000, strongly argued this case. A presentation of these issues was made to the DOE-Nuclear Physics Office in April 2002 by Konrad Gelbke, Brad Sherrill,

and Michael Thoennessen. Partly this was to counter the possibility that Argonne would eventually propose a facility that would be cheaper and based mainly on ISOL rather than PF, arguing that the MSU Coupled Cyclotron Facility was sufficient for the most interesting physics with fast beams. If funding was tight, as seemed likely, this argument could carry significant weight.

- That it had the capability to design and construct a superconducting linear accelerator of unprecedented beam power. This case had to be made in time for a DOE request for proposals that was expected to be issued in 2004. DOE had established mechanisms for obtaining funds for RIA research and development, and MSU took advantage of this funding to finance a variety of projects including development, beginning in 1999, of superconducting radio-frequency cavities and cryomodules that combined several cavities in an acceleration module (see fig. 79).

- That the playing field would be level, so proposals would be judged on their merits. MSU President Peter McPherson worked to ensure that the Michigan congressional delegation was aware of the benefits the new facility would bring to the State of Michigan; was supportive of MSU's RIA/(eventually FRIB) efforts; and would make its support known to the federal agencies. A state contribution of about $100M would support the proposal during the forthcoming competition.

Competitive Advantages

It was obvious that accomplishing all these goals would be difficult. Many felt the odds of success were one in ten or smaller, making it urgent to strengthen the MSU case wherever possible. One attractive possibility was to demonstrate the capability of reaccelerating beams from a particle fragmentation facility. So, during the lengthy FRIB process, in September 2006, NSCL decided to build a reacceleration facility, funded by MSU, as part of its eventual contribution to FRIB. If FRIB was not built, or was not located at MSU, it would also have formed a new direction for research at the NSCL.

The facility, called ReA3 (for ReAcceleration of ions to 3 MeV/nucleon), would use fast nuclei from the A1900 fragment separator, slow them in a gas stopper as shown in figure 80, and then reaccelerate them in a superconducting linear accelerator. ReA3 would provide unique radioactive beams, especially suited for studies of nuclear reactions of interest for astrophysics. It would also serve as a test bed for the superconducting cryomodules that are a major part of FRIB, and eventually serve as part of a reaccelerator for FRIB. Indeed, problems with the ReA3 cryomodules were encountered and were eventually addressed in a small-scale environment that was much more forgiving than the FRIB project would have been. ReA3 is now operating, using a gas stopper built at ANL, and the inverse-cyclotron-based beam stopper,[179] a massive device shown in figure 80, is, in summer 2015, in final stages of construction.

Now that FRIB is being built at MSU and ReA3 is operational, it provides a successful demonstration of the reacceleration of stopped beams that is at the heart of the MSU approach to FRIB.

A second NSCL advantage—its major role in the Joint Institute for Nuclear Astrophysics—initially seemed less important. Although NSCL activities aimed at obtaining JINA began in 1999, at almost the same time as those for FRIB, JINA was not initially conceived as a direct advantage in the competition for FRIB, but eventually it came to play a major role. Many NSCL researchers became active in experiments related to nuclear astrophysics, and it was a field of interest for new faculty hires. The large NSCL effort in that direction and the realization that the reacceleration capability of ReA3 and later FRIB would provide major advantages for studies of reactions important in astrophysics gave MSU an advantage in the competition for FRIB.

■ Building ReA3[180]

ReA3, shown in figure 81, can produce beams with final energies ranging from a minimum of 0.3 MeV/nucleon to a maximum of 3 to 6 MeV/nucleon, depending on the isotope.

Among reaccelerators operating or in progress, ReA3 most closely resembles REX ISOLDE[181] at CERN in Geneva, Switzerland, and the final energies are similar for heavy nuclei. There are, however, two critical differences. First, ReA3 accelerates isotopes produced by projectile fragmentation and slowed in a gas-filled stopper, but REX ISOLDE accelerates isotopes made at rest in the ISOL process. This gives ReA3 access to a different set of radioactive nuclei and to nuclei with shorter lifetimes. Second, the beam from the gas stopper is sufficiently precise that it can be directly and continuously injected into the EBIT, rather than injected in a batch process following an intermediate ion trap as in REX ISOLDE. This removes a limitation on intensity but is challenging to carry out. ReA3 is the first reaccelerator in the world to use this approach.

Construction of ReA3 took advantage of earlier NSCL work on superconducting cavity development. But it was the first linac-based accelerator system of any sort in the laboratory, and it was a challenge to progress through the learning curve for all the new technologies that had to be mastered.

An initial choice to build the accelerator on a platform about ten feet above the laboratory floor

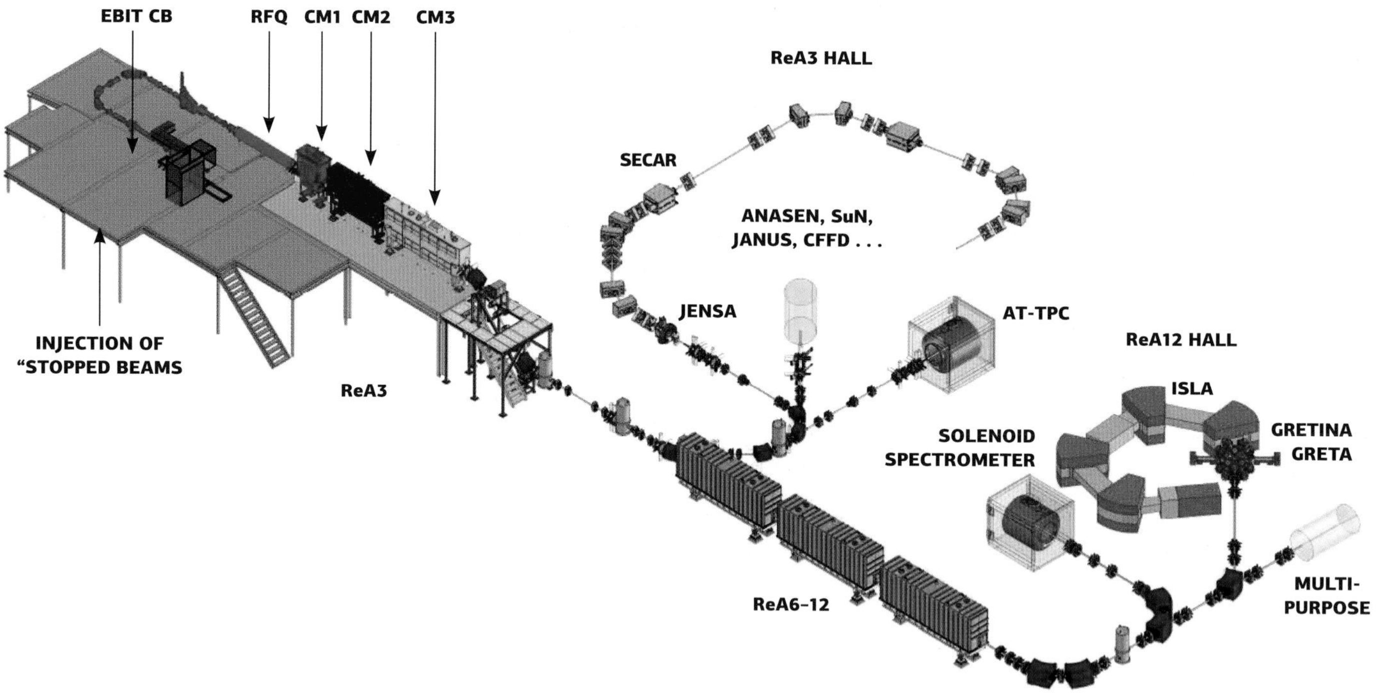

FIGURE 81. The ReA3 system consists of a gas stopping system (see fig. 80), an electron beam ion trap (EBIT) to produce highly charged ions, a room-temperature Radio-Frequency Quadrupole accelerator (RFQ), a linac comprised of fifteen superconducting resonators operating at 80.5 MHz (CM1, CM2, CM3), and a beam-transport system to deliver the beam to the ReA3 experimental hall. The linac can accelerate or decelerate beams from the fixed energy output of the RFQ (600 keV/nucleon) to final energies ranging from 0.3 to 3 MeV/nucleon for heavy secondary ions like uranium. For N = Z nuclei, one can reach 6 MeV/nucleon. The ReA3 experimental area and possible expansions to ReA6 and ReA12, the numerals giving the energies for heavy ions, are shown.

was advantageous in the use of space but introduced an unforeseen problem: vibrations were much larger than a floor-mounted system would have encountered. As a result, the resonators slipped out of tune and operation was impossible. This problem was solved by increasing the rigidity of the platform and by new techniques implemented in the low-level radio-frequency electronics.[182] Another major issue was learning to produce the radio-frequency cavities, with their associated superconducting focusing solenoids, and to do so reproducibly.

By summer 2013, most of these problems had been dealt with. The gas stopper and all parts of the accelerator through the first eight resonators, sufficient for acceleration to 1.55 MeV/nucleon, and the first part of the beam line to the experimental hall had been installed and tested with stable beams from an auxiliary source. The resonators were successfully tested over long periods to gradients well beyond the original requirements. Then, in April 2013, during a first attempt, radioactive beams from the coupled cyclotron and the gas stopper were successfully accelerated, validating the entire design; on August 20, 2014, the first radioactive beam, ^{37}K, was delivered to experimenters.

In November 2014, beam was accelerated through all three modules, signaling the completion of the project and that an experimental program could begin. There remain the challenges of increasing the yields from the gas stopper and the EBIT to obtain anticipated efficiencies of a few percent, but a beam list has been developed, and ReA3 has been added to the call for user proposals.

Since the onset of ReA3 planning, the NSCL user community has expressed strong interest in higher-energy reaccelerated beams and pushed the NSCL to design and develop larger ReA6 and ReA12 accelerators. Unfortunately, funding has not become available for this purpose. It was requested in the 2010 NSCL operating proposal to NSF but was not approved, in spite of strong positive reviews. We remain optimistic that a funding scenario can be developed.

▪ Parallel Development of Nuclear Astrophysics: JINA and JINA-CEE

In June 1999, Sam Austin (MSU) and Michael Wiescher (University of Notre Dame) convened a workshop to discuss research opportunities in nuclear astrophysics. Attendance by 170 researchers reflected the growing interest in this field.

A White Paper based on the conclusions of the workshop, "Opportunities in Nuclear Astrophysics; The Origin of the Elements," was widely distributed. It was a propitious time for such efforts because of the confluence of three technologies:

- accelerators producing intense beams of radioactive nuclei so one can measure the nuclear reaction rates influencing phenomena in the cosmos;

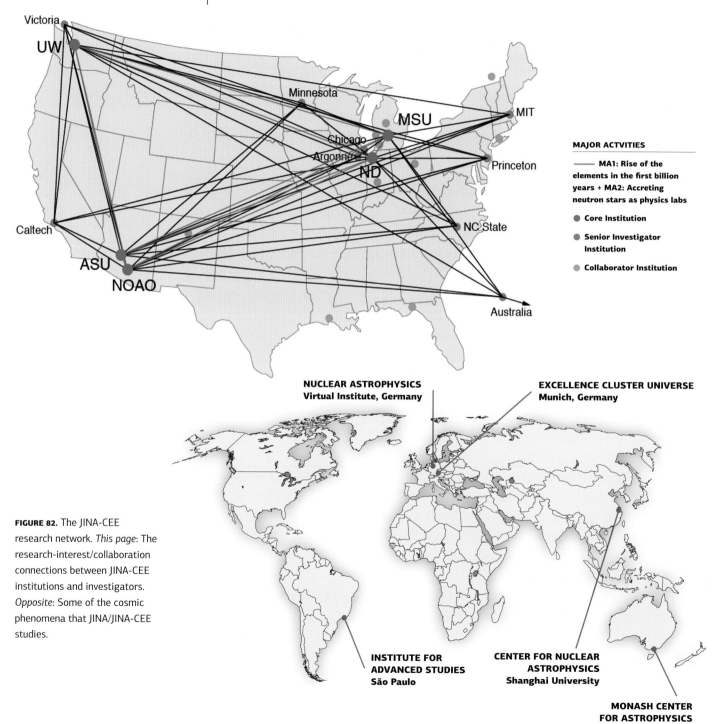

MAJOR ACTVITIES

— MA1: Rise of the elements in the first billion years + MA2: Accreting neutron stars as physics labs

● Core Institution

● Senior Investigator Institution

● Collaborator Institution

NUCLEAR ASTROPHYSICS
Virtual Institute, Germany

EXCELLENCE CLUSTER UNIVERSE
Munich, Germany

INSTITUTE FOR ADVANCED STUDIES
São Paulo

CENTER FOR NUCLEAR ASTROPHYSICS
Shanghai University

MONASH CENTER FOR ASTROPHYSICS

FIGURE 82. The JINA-CEE research network. *This page*: The research-interest/collaboration connections between JINA-CEE institutions and investigators. *Opposite*: Some of the cosmic phenomena that JINA/JINA-CEE studies.

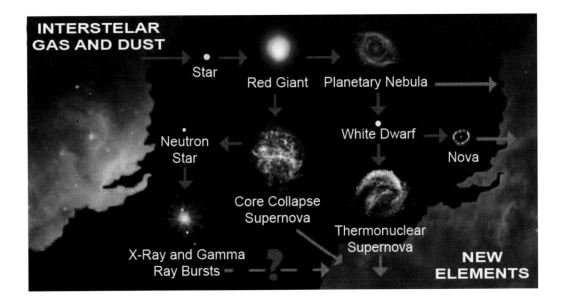

- high-performance computers and associated computer programs that could use these reaction rates to simulate cosmic processes such as supernovae; and
- new and powerful telescopes that could provide observations to test these simulations.

An effort was then made to form a new center that would provide a focus for activities in nuclear astrophysics that had been lacking since the end of the influential program at the California Institute of Technology headed by Nobel Laureate Willy Fowler. Funding this effort was not straightforward. An early MSU proposal to the NSF yielded startup funds. Proposals to form an NSF Physics Frontier Center were prepared and on the second try, a collaboration of the University of Notre Dame (Michael Wiescher PI), MSU (Hendrik Schatz PI), and the University of Chicago (James Truran PI) was funded. Operation of this Joint Institute for Nuclear Astrophysics (JINA) began in 2003 with Michael Wiescher as director and, after a renewal, ended in 2014.

From its creation in 2003, interest grew steadily, and many researchers chose to be associated with JINA. By the end of 2013, there were fifteen Associated Institutions (with Memorandums of Understanding) plus nine participating institutions. Then in 2014, MSU, University of Notre Dame,

Arizona State University, and University of Washington proposed to establish the JINA-Center for the Evolution of the Elements, with MSU's Hendrik Schatz as director. JINA-CEE received funding of $11.4M for a first five-year period. At present, about sixty-five senior investigators in thirty-one organizations are involved in the JINA-CEE research network.[183] This network and some of its research directions are shown in figure 82 and more details are available on the JINA-CEE website, currently www.jinaweb.org.

Most proposals for new rare-isotope-beam facilities worldwide used studies of nuclear astrophysics as a major motivation, and it seemed likely that MSU's position as a prime mover in this direction would prove advantageous.

The RIA/FRIB Competition: Argonne and NSCL/MSU

In April 2002, NSAC formally presented LRP-2002 to DOE, although its recommendation that RIA was NSAC's highest priority for new construction had been known a year earlier. Intense activity followed:

- June 17, 2002, a presentation on *RIA at MSU* to senior DOE officials.[184]
- October 14, 2002, the first meeting of the MSU's RIA Advisory Committee comprised of influential representatives from Michigan business, labor, education, and government, was held at NSCL.
- December 17, 2002, submission of *The Vision for RIA at MSU-Michigan State University's Institutional Plan for Maximizing the Focus and National Impact of RIA at MSU* to DOE-SC Director Ray Orbach at his request.

Activities were also underway at the federal level that would affect RIA and MSU. The DOE published *Facilities for the Future of Science: A 20-Year Outlook*. In this report, RIA was ranked in a tie for third, after ITER (the International Thermonuclear Experimental Reactor) located in France, and UltraScale Scientific Computing Capability—a high enough ranking to compete for funds.

Then, in November 2003, an NSAC Subcommittee headed by Peter Bond of BNL was asked to compare the capabilities of RIA and the somewhat similar FAIR facility at GSI. This was in the context of a request from German authorities for the U.S. to participate in the construction of FAIR and its later use, together with government interest in sharing the cost of expensive facilities internationally, and not building duplicate facilities.

The Subcommittee concluded in its February 23, 2004, report that:

> RIA will provide yields of any element at intensities that are unmatched by any facility, present or currently planned. The RIA and GSI facilities are largely quite distinct in their strengths and are indeed, as the proponents claim, complementary. RIA clearly has a much larger reach as a rare isotope facility, and hence the better facility to address the science associated with rare isotopes.

This cleared the way for future events.

▪ From RIA to FRIB

On October 5, 2004, DOE issued a Draft Request for Proposals (RFP) for RIA. MSU hired consultants familiar with DOE processes and formed an RIA Proposal team. Their first concern was to ensure that DOE's draft RFP did not include requirements that would prevent MSU from competing effectively. The team also began the lengthy process of preparing a persuasive proposal. When DOE cancelled the RFP on February 5, 2005, the team was put on hold.

At this point, MSU turned its attention to designing a less-expensive, alternative facility that would achieve most of the goals of RIA but would have a better chance of being funded. The wisdom of this choice was validated when DOE decided to reduce the budget for RIA.

This effort resulted in plans for an integrated facility based on a superconducting linear accelerator, and in November 2006, MSU presented a detailed conceptual design—*Isotope Science Facility at Michigan State University: Upgrade of the NSCL rare isotope research capabilities*—in MSUCL-1345, known as the Blue Book. The MSU concept was based on a high-power superconducting linear accelerator as a driver,[185] the PF method for production of rare isotope beams, and the possibility of stopping these beams and reaccelerating them. The MSU design took advantage of much of the

NSCL's existing infrastructure for the experiments with fast (PF) beams. Options of locating the facility either on a new, green field site south of MSU or on the present NSCL site were described. The green field site was more expensive and, eventually, was not adopted.

The Tribble Committee, formed by NSAC to provide "Guidance for Implementing the 2002 Long Range Plan," concluded in June 2005, that "RIA remains the highest priority of our field for major new construction." In November, NSCL presented its case for a lower cost (around $500M) facility to the White House Offices of Management and Budget (OMB) and Science and Technology Policy (OSTP).

At about the same time, DOE requested that the National Academy of Sciences review the situation. In response, the Academy's Rare Isotope Science Assessment Committee (RISAC) was formed and first met in December 2005. In February 2006, DOE announced that it would not build RIA, but would build a lower cost (half of RIA cost) reaccelerated exotic beam facility starting in 2011.

RISAC heard presentations from both MSU and Argonne that were consistent with these cost guidelines, with MSU emphasizing fast (PF) beams and Argonne, ISOL beams. RISAC redesignated the reduced RIA as the Facility for Rare Isotope Beams (FRIB) and endorsed the scientific case for FRIB.

■ From FRIB Proposal to FRIB Project

On July 17, 2006, NSAC was charged by DOE/NSF "to perform an evaluation of the scientific 'reach' and technical options for the development of a world-class facility in the United States for rare-isotope beam studies within the funding envelope . . ." just described.

Again, both MSU and ANL presented their ideas to the NSAC Task Force, and in November 2006, MSU published a detailed conceptual design in the Blue Book mentioned above, that made its case to the broader community.

In May 2007, the NSAC Rare Isotope Beam Task Force released its draft report to the LRP writing group and, in August 2007, its final report. Its recommendations were consistent with the Blue Book concept and stated:

> We recommend that DOE and NSF proceed with solicitation of proposals for an FRIB based on the 200 MeV, 400 kW superconducting heavy-ion driver linac at the earliest opportunity. This unique facility

will have outstanding capabilities for fast, stopped, and reaccelerated beams. It will be complementary in reach to other facilities existing and planned, world-wide.

One of the great advantages of the heavy-ion driver/gas stopper/post-accelerator combination is that it allows fast, stopped and reaccelerated beam experiments to be carried out at the same facility. The low-energy nuclear science community has stated repeatedly that it would like to make use of all three capabilities. We see no advantage in mandating the specific mix of these capabilities today; rather, appropriate steps must be taken to ensure that a new FRIB have the strongest and most exciting science program, in the world context, on the day when it starts to operate, ten years from now.

At a meeting of NSAC on December 3, 2007, Under-Secretary Ray Orbach announced that DOE would go forward with a Funding Opportunity Announcement and site selection for a reaccelerated rare-isotope facility but that DOE would not pursue the fast-beam capability. An NSCL response to DOE on December 14, 2007,[186] outlined in some detail the science that fast beams would make possible and pointed out that removing this capability would:

- be against the guidance of the Nuclear Science Advisory Committee and the National Academies of Science;
- severely reduce the scientific reach of FRIB and remove a large part of the scientific justification for FRIB as articulated by the National Research Council's Rare Isotope Science Assessment Committee (RISAC) and by NSAC's Rare Isotope Beams Task Force; and
- relinquish scientific world leadership in the area of rare-isotope science with fast beams where it currently exists in the United States.

In February 2008, the DOE-SC published a Draft Financial Assistance Funding Opportunity Announcement for an FRIB, consistent with the fast beam option, and in May 2008, the final document. Applications were due on July 21, 2008.

There followed a flurry of activity to produce a proposal that was consistent with DOE guidelines, met DOE criteria, and made a strong scientific and technical case. This involved a small team, working in seclusion off-site at MSU in a student residence, Akers Hall. Both MSU and ANL submitted "compliant applications." These documents are, apparently, not available in the public record at the time of this writing.

Lansing State Journal

FRIDAY, DECEMBER 12, 2008   THE POWER OF KNOWING SINCE 1855   ★ WWW.LSJ.COM   75¢

Keanu is Klaatu
Reeves is well-cast as alien visitor in 'The Day the Earth Stood Still' remake
Life • 1D

No Doak
Walker Award for top RB eludes MSU's Ringer
PAGE 1C

$550M WINNER

MSU LANDS NUCLEAR RESEARCH FACILITY

» **'GAME CHANGER':** Hollister says project changes mid-Michigan's economic future

» **ECONOMIC BOOST:** Project expected to bring $1 billion in new economic activity over 20 years

MATTHEW MILLER
mrmiller@lsj.com

Hope arrived in Michigan on Thursday on the backs of a billion unstable atomic particles.

The U.S. Department of Energy announced it had awarded Michigan State University a $550 million nuclear physics research project called the Facility for Rare Isotope Beams.

And state and local leaders began to talk as if the much-touted new economy had begun to arrive.

"By getting this designation, we can say we have shed our Rust Belt image," said former Lansing Mayor David Hollister, who now heads the regional economic development agency Prima Civitas. "This will bring about transformative innovation. It is a game changer for our state."

Multiple benefits

Estimates say the project will bring $1 billion in new economic activity and $187 million in tax revenue over 20 years; 180 new jobs for scientists and facility staff; 5,800 one-year construction jobs; and 220 spin-off jobs.

"You talk about a rare isotope accelerator, this is going to be a business accelerator," said Lansing Mayor Virg Bernero. "It is going to be an economic development accelerator.

"If properly marketed — and it will be — this is great, great news in terms of that high-tech sector that we have been working on."

For MSU, it means a place

See **MSU** | Page 3A

All smiles: Michael Thoennessen, professor and associate director for education at the National Superconducting Cyclotron Laboratory at Michigan State University, shares the happy reaction to the news that the Facility for Rare Isotope Beams is coming to East Lansing.
ROD SANFORD/Lansing State Journal

❝It is Christmas coming early to some extent. (The Facility for Rare Isotope Beams) is the future of our laboratory.❞
Zachary Constan, 35, of Lansing, outreach coordinator for the National Superconducting Cyclotron Laboratory

❝When you look at the fact that the politics are such that the new president comes from the state where their competitor was, the fact of the matter is this was not about politics, it was about merit.❞
Sen. Debbie Stabenow, D-Lansing

Facility for Rare Isotope Beams

Construction is to begin in 2009 on the Facility for Rare Isotope Beams, or FRIB, which will be designed and built on the Michigan State University campus. It is to be completed by 2017.

1 Law College (existing)
2 National Superconducting Cyclotron Laboratory (existing)
3 Wharton Center (existing)
4 Office and experimental area expansion (to be completed by 2009)
5 FRIB expansion (to be completed by 2017)
6 Linear accelerator tunnel

E. Shaw Lane

Source: Michigan State University   Martha Thierry/Detroit Free Press

WHAT IT MEANS

JOBS

"It's going up a lot of work for local industry," Terry Grimm said. His Lansing firm, Niowave, manufactures parts for superconducting particle accelerators like the one that will be built at MSU, and employs 33 people. To do the work for the FRIB, Grimm said, would require "150 to 200 employees," most of them engineers and skilled machinists.

MORE JOBS

The announcement was good news for unions that represent construction workers. "It will be good to get our people back to work," said Dan Minton, secretary-treasurer for Laborers Union Local 499. He said 35 percent of his 1,500 members are currently unemployed.

HIGH SCIENCE

The FRIB will draw leading scientists from around the globe and could spark the creation and growth of high-tech companies in the area, said Scott Watkins, senior consultant at Anderson Economic Group in East Lansing.

MICHIGAN'S MORALE

"This is a huge victory ...," Gov. Jennifer Granholm said. "Bringing this facility here signals to the rest of the nation that if you want to grow your business in a state that is on the cutting edge of science and technology – Michigan is the state."

FIGURE 83. Headline in the *Lansing State Journal* on December 12, 2008.

The proposals were reviewed by a large team of technical advisors to DOE and by federal officials. MSU and ANL were both invited to make presentations on October 14 and 15 in Rockville, Maryland. MSU President Lou Anna K. Simon and Vice President Fred Poston led the MSU team, accompanied by other high MSU administration officials and members of the proposed MSU FRIB team: Konrad Gelbke, MSU FRIB Laboratory Director; Thomas Glasmacher, Project Manager; Richard York, Technical Director; Brad Sherrill, Chief Scientist; and Georg Bollen, Experimental Systems Division Director. MSU faculty members Paul Mantica and Hendrik Schatz gave talks on the proposed science program.

On October 20, the technical advisors to DOE and the federal officials came to MSU for a site visit and additional presentations given in MSU's Henry Center.

Then on December 11, 2008, DOE announced that:

> Michigan State University (MSU) in East Lansing, Michigan, has been selected to design and establish the Facility for Rare Isotope Beams (FRIB) . . . The selection is subject to the successful negotiation of a Cooperative Agreement with MSU and a National Environmental Policy Act (NEPA) review of the proposed site. Funding is subject to annual appropriations by Congress.

Figure 83 shows that this event was received with great enthusiasm at the NSCL, at MSU and in the community.

The cooperative agreement was signed by the DOE and MSU on June 8, 2009, marking the beginning of FRIB. A dedication ceremony, *Rare Isotope Beams for the Twenty-First Century*, was held on June 10, 2009. It celebrated this major expansion of the NSCL and looked ahead to the Facility for Rare Isotope Beams—FRIB at MSU. This was, indeed, an event to celebrate the result of the hard work, persistence, and resilience of many people.

There were significant delays in federal funding owing to the Budget Control Act of 2011 (sequestration), which led to uncertainty and frustration. But it also gave the technical team time to perfect their plans and designs. Finally, FRIB became a project when, following a two-day review by the U.S. DOE-SC from June 4 to 5, 2013, DOE-SC announced on August 5 that it approved Critical Decision 2/3A: the project cost and schedule (baseline), and civil construction and long-lead procurements. Civil construction was scheduled to, and did, start in 2014.

HOW THE CASE WAS MADE

From the beginning, in 1999, the MSU Administration: President McPherson, Provost and then-President Simon, and the MSU Board of Trustees, had played a crucial role in supporting initial plans for the Rare Isotope Accelerator (RIA) and later for the less-expensive FRIB, even when success seemed improbable. They marshalled the support of Michigan's Governor and Congressional delegation and of its business and community leaders. And they provided and helped secure critically needed funding support for technical R&D, conceptual development of the project, and for the required space and facilities to make progress possible.

President Simon played a particularly important, strong, and supportive role throughout the competition. She led the MSU team for the presentation to the DOE selection committee in Rockville, Maryland, October 14–15, 2008, and she was key in the site visit to MSU October 20–21, 2008. Her personal commitment, intimate knowledge, and determined support of the project were important in persuading the DOE and its technical advisors that MSU would provide the necessary support, in partnership with the DOE, for the eventual success of the FRIB project.

DOE-SC conducts regular reviews of the FRIB Project that provide vital oversight and approvals as it develops from stage to stage (see Appendix A).

The Scale of Change: Two Bird's-Eye Views

As the laboratory has evolved and grown, its environment has evolved as has its understanding of what one must do to make growth possible. One can see the growth of the laboratory most clearly in its buildings. Figure 84 shows how growth from the original center has resulted in an extraordinary compact layout. This has great pluses in enhancing interactions among those with different interests but also has associated space limitations. Figure 1 shows the present layout including the additional FRIB laboratory. The design of FRIB presented to DOE involved a linear accelerator that extended across Shaw Lane. However, for both technical and cost reasons, that design evolved and now has

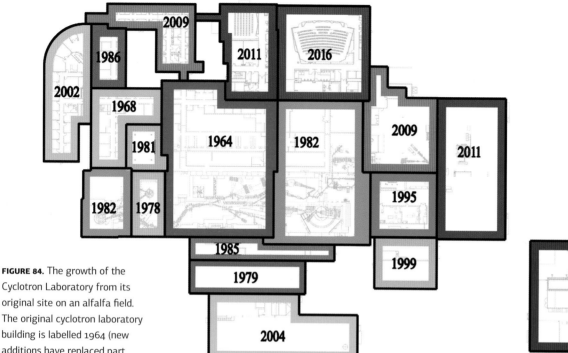

FIGURE 84. The growth of the Cyclotron Laboratory from its original site on an alfalfa field. The original cyclotron laboratory building is labelled 1964 (new additions have replaced part of it), and the 1968 addition provided laboratory space and offices for theorists. There was no additional construction in the next ten years, but in 1978 a small building was added to house the new K500 Cyclotron, and in 1982 there was a major building expansion for the Phase II project, also now partly replaced. Several small additions in the 1990s were mostly for various coupled cyclotron activities. In the 2000s, most construction was related to the activities to design and build FRIB.

what some call a paper-clip shape that fits on the Cyclotron Laboratory site. That there was space on the site for all these expansions owes to the early foresight of building the K50 in a new area of campus, rather than next to the Physics-Astronomy building.

Understanding what one must do to make growth possible can be seen in figure 85. An examination of the time scales—from idea, to completion of construction, to completion of science with the facility—is instructive. Obtaining a new accelerator system takes around ten years and sometimes longer, and a program of forefront science lasts from ten to twenty years. To maintain a forefront research program, one then has to begin planning for a new accelerator relatively early in the life of the present accelerator, when it is probably in its most productive years. This is typically a difficult decision: diversion of resources to planning for the future affects the ongoing research program. But such foresight is crucial for long-term success.

■ Entering a New Era

This brings our account full circle, but it is not an ending, only the next beginning. As with the MSU Cyclotron Laboratory's rise "up from nothing" in 1958, to the world's most precise cyclotron—the K50—and now to FRIB, many uncertainties and many years of hard work remain before the completion of FRIB.

During the next seven years, the laboratory will be a beehive of activity as the many technical components of FRIB are assembled and installed: the ion source front end, driver linac, beam delivery system, isotope production target, shielding, fragment separator, experimental areas, and all their many subcomponents. Surely there will be difficulties to address and technical challenges to meet. To overcome them will require the combination of expertise, foresight, technical agility, persistence, and teamwork that have served the laboratory so well in the past.

When FRIB is complete, in 2022 or perhaps somewhat earlier, the MSU Cyclotron Laboratory

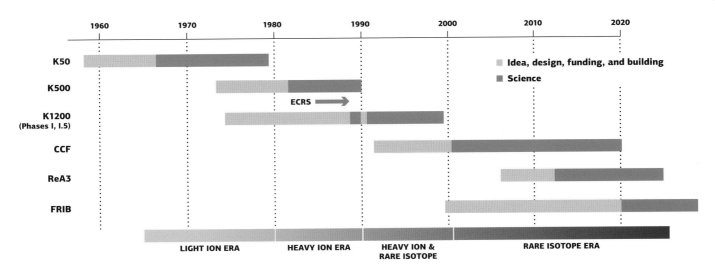

FIGURE 85. Time scales for construction and use of the MSU Accelerators. The rows describe the various MSU accelerators and, along the bottom, the nature of the research effort with each accelerator.

LABORATORY FUNDING: PAST AND FUTURE

The scale of what the Cyclotron Laboratory has undertaken in the past may help us understand the challenge that awaits us, as physical construction of FRIB goes forward. We use expended funds as a rough measure of past effort as shown in these charts.[187] All funding is expressed in 2014 dollars. In the years before 1980, nearly all funding was from the NSF, and NSF funding dominated what is enumerated here throughout this period. Neither MSU support nor DOE support directly related to FRIB is included. An extrapolation of NSF funding for CCF operation up to 2020 is also shown.

YEAR	FUNDING*	YEAR	FUNDING*	YEAR	FUNDING*
1962	$5.5M	1980	$5.5M	2000	$22.7M
1965	$3.2M	1985	$18.3M	2005	$25.6M
1970	$4.6M	1990	$17.0M	2008	$23.8M
1975	$6.2M	1995	$18.3M	*Non-FRIB, in 2014 $	

ERA	YEARS	FUNDING (2014 $)
Total, to date	1962–2008	$664.5M
Total, K50	1962–1979	$101.3M
Total, K500	1980–1987	$143.3M
Total, K1200	1988–1999	$211.0M
Total, CCF	2000–2008	$208.9M
Total, CCF (extrapolated)	2009–2020	$300.0M

Two messages can be gleaned from these figures. First, that the Cyclotron Laboratory has been a large operation for some time. Second, that although FRIB is the largest project the laboratory has undertaken by a significant factor, on a per year basis, average funding is larger by perhaps a factor of 2.5 to 3 from what we have been spending. This change is similar to that experienced going from the K50 era to the superconducting cyclotron era as shown. So FRIB is a challenge, but not an overly worrisome one.

will complete its most impressive reinvention to date and will play a preeminent role in advancing the frontiers of nuclear science in the twenty-first century.

While FRIB is under construction, the MSU/FRIB Laboratory, née Cyclotron Laboratory, remains a complex and interlocking organization, with two distinct sets of tasks. One is to build FRIB, a construction project larger by a factor of ten than previous NSCL construction projects. As such, it is subject to close oversight by DOE, with an exhaustive series of advisory committee recommendations and reviews. It has firm deadlines and a formal, highly disciplined project-management structure. Thomas Glasmacher as project director is responsible for all aspects of the FRIB project.

A second task is for NSCL to perform research of the highest quality with the Coupled Cyclotron Facility and to educate graduate and post-graduate students. NSCL and FRIB leadership must also develop the infrastructure necessary for research to begin, using FRIB, on Day One after its completion. Both leaders and faculty must remain deeply involved with the international nuclear science community, the users of the future FRIB, as the FRIB laboratory prepares to integrate users into research in the FRIB era.

As it has always been, the laboratory structure is fluid and in an evolving state of high complexity, but now the complexity is greater. We have experienced changes over the years in the way the laboratory operates and these will continue, reflecting changing circumstances. As an obvious example, many companies and other laboratories are involved in FRIB and NSCL construction projects, and contracts must be carefully managed. Because there are many sources of funding involved, and a single individual may receive support from several of them and for several different projects, it is necessary to fastidiously account for the time spent on each project, a much more complex task than in the past. To make sure this is accomplished, the FRIB business support organization performs procurement, finance, human resources, and many other functions.

FRIB expenditures will be far larger than NSCL expenditures, so the FRIB business organization will also handle parallel NSCL activities. There are overlaps of function in other activities as well, and these are sometimes integrated and sometimes kept separate as seems most efficient in each case. Other changes will follow as we seek to optimize laboratory operation.

Two facts will govern our choices and actions: (1) In the near term, the crucial priority is to build FRIB while maintaining efficient operation of the Coupled Cyclotrons, and (2) in the future, the crucial priority will be to do important science with FRIB beginning on Day One, while maintaining efficient operation of FRIB and improving its capabilities.

This evolving dichotomy is a challenge to all of us. Moreover, the cultures of a large construction project and of an academic research organization are different, and each may operate more efficiently with a different set of procedures. It will be important to respect these differences as we are passing from a primarily research activity to a major construction project, and then, later, back again to a larger research environment.

The NSCL and FRIB will have to face and resolve many operational and interaction issues as FRIB proceeds. The history of the laboratory shows that an overall unity of purpose and forward-looking goals will lead to long-term success, the promise of successful completion of FRIB and a bright future for nuclear science.

Looking Back: Building upon Increasing Strength

During this account I have tried to delineate the characteristics of the Cyclotron Laboratory that led to its success by following its journey "up from nothing" in 1958 to FRIB. These characteristics, which constitute its culture, together with its research prowess are why it was ultimately entrusted to establish FRIB, the nation's premier nuclear science facility for the twenty-first century.

When the Cyclotron Laboratory was founded in 1958, it was smaller and less known than most university-based laboratories. Yet, while many established laboratories at well-known institutions declined and even disappeared, MSU's nuclear science program grew and is now generally regarded as the best of university-based programs, and better than most national laboratory programs.

■ A Culture of Agility, Persistence, and Performance

The culture, which developed over time, from 1958 to the present, rests upon major strengths that comprise its foundation.

CULTURE: WHY THE LABORATORY THRIVED

- Support from MSU
- A tradition of build-it-yourself
- Faculty attitude and quality
- Senior leadership
- A tradition of looking ahead

- Close connection with theorists interested in the experimental program
- Outstanding students and colleagues
- Luck

SUPPORT FROM MSU

Support for the Cyclotron Laboratory in the late 1950s was crucial to its initial success, and included: startup funding of the nascent accelerator group, persistent assistance with persuading agency personnel to fund K50 proposals, funding of a building to house the K50 cyclotron, and preventing bureaucratic concerns from handicapping the project. This kind of assistance has continued throughout the life of the laboratory. MSU has been eager to bolster its reputation as a growing research university, and a new Cyclotron Laboratory expansion was often the major new development on campus, promising the greatest potential for increased reputation and significant federal funding. As a result, the laboratory received MSU monetary support to fund new developments and infrastructure that would leverage additional grant funding. It also won support for establishing its own in-house purchasing system and new personnel systems that maximized its agility and responsiveness.

There were often tensions around the extent to which MSU funding might be required to support operation of the facility. The NSCL and MSU viewed these funds as research facilitation funds, dedicated to support new infrastructure and speculative developments that likely would pay future scientific dividends for the laboratory. But when federal funds were tight, NSF sometimes limited increases in funding levels to less than inflation or reduced funds. The laboratory then had two choices: to use any available MSU funds to support operation, or to cut operating expenses and the number of experimental hours available. The latter was difficult, given that the demand for the facility was oversubscribed by a factor of two, and that the university and laboratory shared a strong desire to retain experienced technical staff who would be needed when funding levels recovered.

Once it became a national user facility, the NSCL concluded that it was not only disadvantageous for the future development of the laboratory, but simply not ethical, to use MSU funds to support operations, and thereby research at other institutions. The NSCL then proposed to limit the hours of operation to what was supported by NSF funding. This approach was generally accepted by NSF and the user community. In certain cases, however, the most conspicuous being the GRETINA[188] campaign in 2012–13, the laboratory concluded that the scientific case was sufficiently strong to warrant additional operation funding from MSU. These funds served the broader national interest in establishing U.S. dominance on the international scene in the rapidly evolving use of gamma ray tracking.

During the development of FRIB, MSU advanced funds to support planning and proposal expenses and building construction—including three office towers, two experimental areas, and a superconducting radio frequency assembly building, so as to provide space for expanding personnel and activities. The availability of MSU funds, with a fiscal year different from federal funds, has also made it possible to smooth out bumps in NSF funding and prevent layoffs and reassignments that would otherwise have disrupted the operation of the laboratory.

A TRADITION OF "BUILD-IT-YOURSELF"

The laboratory tradition of building experimental apparatus and developing forefront tools for experiments in house can be traced back to the reasons for the original choice made in 1956 to build a cyclotron rather than buy a Van de Graaff from a manufacturer: "that those who build the machine will have an intimate knowledge of its properties and that this will lead to easier maintenance and more imaginative research."[189] This was borne out in the development of the high-resolution capability of the K50 cyclotron.

Later, this approach developed into a general philosophy, sometimes stated in proposals to the NSF: decide what your experiment needs; design the apparatus that meets that need; and then build it in house, unless it happens to be available as a less expensive, off-the-shelf item (rarely), or the size or special apparatus required makes local building impossible.

This approach avoids the compromises of settling for what you can buy with the available funds. As an example, although most laboratories would buy the bending and focusing magnets for a beam line, we always built our own—first room temperature magnets for the K50 and then superconducting magnets for later accelerators. An important example was the A1200 fragment separator, which was conceived at the last minute and constructed by modifying magnets that had been made for a standard beam line. This general philosophy has also had an influence on hiring decisions; we usually hired faculty who had an interest and capability in apparatus construction. These decisions are responsible for the remarkable list of contributions to the apparatus of nuclear science that are listed in Appendix B.

FACULTY ATTITUDE AND QUALITY

As was discussed during the description of the K50, the initial faculty group was involved in all aspects of laboratory operations, experiments, and governance. In addition, there were no groups, rather a set of evolving collaborations, involving different people at different times. This, and the feeling that a laboratory was being built from scratch, enhanced camaraderie and a sense of shared purpose that were strongly positive for laboratory development. Faculty and senior staff thought not only of their own success, but at least equally, that of the laboratory as a whole. This attitude also led to the success of attempts to recruit high-quality faculty members who might not normally have considered such a fledgling institution.

This no-groups, evolving-collaborations approach prevailed until the K500 cyclotron began operation with heavy-ion beams. Some of these experiments involved complex detectors, eventually the 4π Array and the Miniball, that generated large amounts of data requiring time-consuming analysis. As a result, it became difficult for post-docs to be involved in several different experiments, as had been common in the past. Dedicated groups were needed to effectively manage time and focus.

That argument carried the day, although the attitude of Sam Austin, as Research Director and Director, was to resist the formation of groups except when absolutely necessary. He had served on NSF committees that advised closing accelerators at many universities. In some cases, the problem was that a dominant researcher had formed a large group that took a major part of the grant resources but was not sufficiently productive. More commonly, multiple groups received similar proportions of grant resources but were not similarly productive. These closures were taken to support the rotating group model that had been the early mode of K50 operation. The value of this model was that it tended to focus resources on the most interesting problems and the most active groups. It also resisted a group-not-lab orientation and its dangers.

As time passed, however, the number of faculty grew large enough and of sufficiently high quality that most of the dangers of groups were counterbalanced by their advantages in efficiency.

The faculty has remained strongly interested in mixed collaborations and strongly supported efforts for lab improvement, sometimes even at a cost to their individual efforts. The general atmosphere now, however, is probably less collegial than it had been, and this is perhaps inevitable, given the growth of the laboratory, now having well over 500 members. It is also abetted by the ability, provided by computerization of experiment preparation and analysis, to work at home, rather than return to work at the laboratory at night. There is, of course, the tendency to romanticize the "good old days." Rex Morin, Executive Director of the laboratory for many years, told me that every generation of laboratory researchers spoke to him of how good it had been in the past. In any case, a lab-first attitude remains, and maintaining it is an important goal.

SENIOR LEADERSHIP

The Cyclotron Laboratory has been fortunate in having senior leadership with the right management characteristics for the tasks at hand and the ability to grow and advance over time. As the laboratory rose up from nothing, the goals of the leaders, the milieu in which they worked, and the development requirements of the laboratory differed. These evolving situations demanded and rewarded people with different skill sets. Fortunately, individuals were available who possessed the appropriate talents.

The four individuals who spent significant periods as directors all had strong research interests with which they made their reputations: Henry Blosser as a builder of new generations of precision cyclotrons and superconducting cyclotrons; Sam Austin in the study of the structure of nuclei and of nuclear reactions important for the evolution of the cosmos; Konrad Gelbke in the study of interactions of heavy ions and their implications for the nature of nuclear material; and Thomas Glasmacher for exploring the structure of rare isotopes by studying the emitted gamma rays. They all thought of their administrative duties as worth doing well so they and future Cyclotron Laboratory researchers could continue to do first-quality science. They also brought to the position different strengths and skills that turned out to be appropriate to the times in which they served. Blosser and Austin returned to science after their director duties.

- HENRY BLOSSER, as founding Director, was in a special situation. He needed and had the personality traits that were necessary to taking the first step in building a laboratory: an impatience with any delay and bureaucracy that would delay this process and the willingness to force his way forward. He combined this with care in building accelerators and was willing to face delay and cost overruns to do things properly. He was also extraordinarily creative in dealing with the inevitable setbacks that occur in any forefront construction project. He was alert to new accelerator opportunities and aimed to keep MSU at the accelerator forefront of increasing capability. At the same time, he recognized that he did not have the background to guide the nuclear science program, and after using care to hire quality researchers, he, in most cases, let them follow their instincts, with high-quality results.

- SAM AUSTIN was co-director and director at a time of different emphases. His priority was to make sure that the new generations of superconducting cyclotrons succeeded in producing quality science. This involved reducing reliability problems with the K500 and attracting and growing a user community in spite of unreliable accelerator performance. He also had to establish an efficient management infrastructure to serve this community. For this purpose, Austin's involvement in the American Physical Society and many other nuclear science community activities gave him the credibility and relationships to make the necessary decisions. The setup of user activities that were put in place still serves the Cyclotron Laboratory well. He also succeeded in nurturing a strong internal research program and recruiting the next generation of MSU/NSCL scientists, with the result that the NSCL not only survived but prospered.

- KONRAD GELBKE faced a still different environment. The NSCL was performing well, but NSF funding was tight and the laboratory had to defend its status and ensure its continuation. Moving—immediately—to a next-generation accelerator system seemed necessary, and Gelbke led the effort to develop, fund, and build the present CCF. His greatest contribution to MSU and the Cyclotron Laboratory was leading the successful effort to establish FRIB at MSU. Most laboratory members felt, initially, that the odds of success were small, perhaps one-in-ten at best, but Gelbke persuaded laboratory members to make this a priority while continuing to produce quality science with the CCF. He played a major part in organizing support within MSU, the state of Michigan, and the national nuclear science community. With tightening budgets, and the billion-dollar scale of the new accelerator, the entire process was highly competitive and political. Fortunately, Gelbke had the ability and willingness to operate very effectively on this stage.

- THOMAS GLASMACHER, as FRIB Laboratory Director and Project Director of the $730M FRIB Project, faces the challenge of building a facility almost a factor of ten larger than the laboratory's Phase II Project and doing so while the NSCL maintains a forefront program of research and education. He led the team that prepared the winning FRIB proposal to the DOE, and built the FRIB Project team and led it through the various DOE reviews and into construction. In all these activities he showed exceptional drive and skill. He has had to build a new culture at the Laboratory that reflects the needs of dealing with a project-oriented federal agency and assigns the highest priory to construction rather than research. His challenge at present differs from those of previous Directors, because the goal for the foreseeable future is determined. He has shown that he possesses the skills to direct the construction of FRIB, to bring it to completion by June 2022 or perhaps earlier, and to initiate a research program using its unique beams of rare isotopes.

A TRADITION OF LOOKING AHEAD

This was, and is, a characteristic of the Directors, Associate Directors, faculty of the Cyclotron Laboratory, and also of MSU's leadership. Identifying, building, and taking advantage of new accelerators and new physics on the leading edge of nuclear science has become a tradition.

As this history makes clear, both leaders and faculty have endorsed this tradition. Flexibility in adapting to new research areas rather than sticking with areas of diminishing interest was assumed. At the beginning, the new faculty moved quickly into science with a cyclotron rather than Van de Graaffs, doing new types of experiments at higher energies. Then, when the K500 was built, there was a move to heavy-ion science and studies of the properties of the bulk nuclear material. And still again, when the K1200-based facilities began operating, they moved to studies with radioactive beams, exploring the edges of the known nuclear chart. Once changing with the times has been done on a few occasions, it becomes a tradition and a strength. The last change, to FRIB, reflects the power of this characteristic of the Cyclotron Laboratory.

A CLOSE CONNECTION WITH THEORISTS INTERESTED IN THE EXPERIMENTAL PROGRAM

Some of the earliest Cyclotron Laboratory hires, even before there was a cyclotron, were nuclear theorists who had an interest in interacting with experimentalists. This has become another laboratory tradition, and has helped greatly in the planning and interpretation of experiments and in increasing the impact of experimental results. It is a characteristic often remarked upon by reviewers and visitors to the laboratory, and is commonly seen in the author list of mainly experimental papers. It has sometimes affected accelerator design choices: George Bertsch's qualitative arguments about the energy necessary to completely overlap the momentum distributions of two heavy nuclei was a principal reason for setting the energy of the coupled cyclotron facility at 200 MeV/nucleon. At present, most of the theoretical group and their students work on problems upon which MSU facilities can have an impact.

OUTSTANDING STUDENTS AND UNIVERSITY-WIDE COLLEAGUES

The Cyclotron Laboratory's educational function, broadly, is to educate the next generation of nuclear scientists. It includes all who have spent time at the laboratory: graduate and undergraduate students, postdoctoral scientists, and former laboratory members who moved on to other high-level scientific or administrative positions.

All of these "students" have, over the years, added great value to the laboratory's growth and quality. Their energy, curiosity, and creativity keep our faculty and staff continuously thinking ahead and working to stay at the forefront of their fields.

As a bonus, we are surrounded by a university community that not only stimulates thinking but cross-pollinates the laboratory's awareness of possibilities for research and teaching with research and scholarship from other disciplines.

LUCK: BEING IN THE RIGHT PLACE AT THE RIGHT TIME, RECOGNIZING IT, AND TAKING ADVANTAGE OF IT

For those who have read this account, it is clear that luck plays a role. But, as usual, it is also true that luck favors those who react creatively to the difficulties of bad luck and the opportunities of good luck. A creative reaction to the bad luck of unreliable K500 performance, and difficulties with the construction of the K1200, and the good luck that ECR sources were invented at the right time, made ECR + K1200 operation successful without coupling the two cyclotrons. The project would eventually have succeeded without this realization, but it would have been delayed and more costly.

These characteristics have emerged and converged to create the culture of the MSU Cyclotron Laboratory and its successive generations and will, I believe, play a continuing role in its future evolution.

■ A Leading-Edge Program of Laboratory Research

In much of this account, we have focused on the construction of the tools of nuclear science and the story of how they were built and evolved. It is, however, an expectation and obligation of a research institution that it provide a product that justifies its funding. In the case of the Cyclotron Laboratory, that product is, mainly, twofold: basic research into the properties of nuclei, and the concurrent education of young scientists. If the laboratory is to be successful, its research must continue to address the most interesting forefront problems while preparing the next generation of nuclear science researchers for the nation (see fig. 86). Secondarily, it attempts to foster applications of its research and technical innovations. For more on Technical Outreach, see Appendix B.

It is impossible to summarize in any detail the great variety of nuclear science, experimental and theoretical, that has been done at the Cyclotron Laboratory over the past fifty years. But I hope to give some sense of the general way in which the laboratory's focus has evolved with time. Partly, this is based upon the Cyclotron Laboratory's description of its most important accomplishments in its proposals for federal support. Each of the following sections includes an introduction to relevant concepts.

We begin with a description of the basic research in nuclear science, and later describe research in cross-disciplinary areas and in applied isotope science. For some of the terminology, it will be useful to refer back to Chapter 5.

THE K50 ERA—1965-79

In the years after WWII, the compound nucleus model dominated our understanding of nuclear reactions and how to use them to obtain information about the nucleus. In this picture, a projectile incident on a target forms a compound nucleus, which has a long life, much longer than the time it would take the projectile to traverse the nuclear target. A proton bombarding a ^{12}C nucleus would,

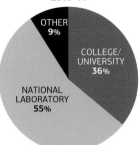

**GRADUATE STUDENTS
2010–14**

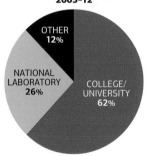

**RESEARCH ASSOCIATES
2005–12**

FIGURE 86. Where graduate students and research associates go after leaving MSU. The large fractions that spend the next stage of their careers at universities and in national laboratories is a testimony to the quality of their training.

MEASURING RESEARCH SUCCESS

A gross measurement of overall research activity is measured by the number of papers published in journals. After five years of operation in 1970, this had reached a substantial level, around twenty-five per year; from 1980–2000 it remained at a high level, around sixty papers a year; then as the NSCL became a truly forefront facility in physics with radioactive beams, the number increased dramatically, and by 2012 had reached 170, with many papers (twenty-nine) appearing *Physical Review Letters* (PRL), the most prestigious journal for nuclear physics. Two or three papers a year also now appear in *Nature* or *Science*, high-visibility journals with broad science coverage that reach a more diverse audience.

for example, form a ^{13}N compound nucleus in a highly excited state. This system eventually decays and its products are observed; one typically obtains information about the compound system (^{13}N).

As the projectile energy increases, it becomes less and less probable to form a compound nucleus. With the high energies available from the K50, the emphasis then shifted toward direct reactions, which take place quickly, in about the time it takes for the projectile to cross the nucleus, perhaps 10^{-21} sec. The K50 together with its high-resolution detection systems was ideal for the study of these reactions and yielded information about the target nucleus and its neighbors.

Many types of reaction were studied. From inelastic scattering of protons we learned how the interaction between protons and neutrons is modified inside a nucleus, and to what extent a nucleus deviates from a spherical shape. We also learned about how the nucleus vibrates, in a breathing mode or through various ellipsoidal shapes, and about nuclear matter and how hard it is to compress. From transfer reactions we learned how and in what orbits a nucleon in a nucleus moves. As we progressed through this era, the development of the Enge spectrograph made possible measurements with very high resolution and opened the range of experiments to the whole periodic table, not just light nuclei.

A beam swinger made possible a new technique for studying neutrons, with excellent resolution. From elastic scattering of neutrons we learned how high-energy neutrons scatter from nuclei and how different are the interactions of two neutrons compared to two protons. We learned that the strength of weak interactions that lead to the decay of radioactive nuclei could be determined from charge exchange reactions; and we discovered the long-sought giant Gamow-Teller resonance that involves oscillations of nucleon spins in a nucleus.

The role that nuclear reactions play in the evolution of stars and in other cosmic phenomena became, and would remain, an important part of the Laboratory program. From the study of spallation reactions induced by protons and alpha particles on C, N, O targets, we developed a model for the origin of the elements lighter than carbon; these results led to a limit of the mass density of the Universe and the tentative conclusion it would expand forever. From a series of experiments on the Hoyle State in ^{12}C, we obtained a more accurate rate for the triple alpha reaction in stars that changed our expectation for the cosmic abundance of ^{12}C, a major constituent of living organisms.

By the end of the K50 era, precision nuclear structure studies had become the hallmark of the Laboratory; it was the world leader for such studies.

THE K500 ERA—1982–90

After the K50 was closed in 1979, new faculty members with experience using heavy ions and other laboratory members did experiments with heavy ions at other laboratories, especially at the Lawrence Berkeley National Laboratory (LBNL). Similar studies continued at the laboratory when the K500 began operation in 1982 and provided unique beams of medium- and high-energy heavy ions. Many experiments involved the nature of nucleus-nucleus (N-N) collisions; from such studies the detailed quantum structure of the interacting nuclei was sometimes used to select the events to be studied but was itself of secondary interest. Other experiments took advantage of the quantum properties of heavy ions to study new aspects of nuclear structure. There was a rough balance of the accelerator time devoted to these two types of experiments.

- THERMODYNAMIC QUESTIONS were central in the study of nucleus-nucleus collisions and the particles emitted from them: what is the nature of the emitting system and does it come to equilibrium so one can define a temperature? We developed different ways of measuring temperature and studied differences obtained with differing techniques—these temperatures are perhaps the highest seen in nature, around 100 billion degrees. It was found that high-energy (>20 MeV) gamma rays were emitted in nucleus-nucleus collisions and probably came from interactions of individual neutrons and protons early in the collision; that emission of light ions in the plane of reactions came from a rotating di-nuclear complex; that nuclei multi-fragment, emitting up to five intermediate mass fragments; and that pi mesons were emitted in nucleus-nucleus collisions, requiring that a large fraction of the incident projectiles' energy is concentrated on a single nucleon in the target nucleus.

- STUDIES OF NUCLEAR STRUCTURE had two main foci: the use of heavy-ion reactions as probes of weak-interaction strength and studies of the production and properties of exotic nuclei and neutron-rich nuclei. Most of these measurements also had connections with nuclear astrophysics. We showed that the (^{12}C, ^{12}N) and (^{6}Li, ^{6}He) reactions provide reliable values of weak-interaction strength with better resolution and sensitivity than nucleon probes, and that the former reaction provides formation needed for theoretical models describing core collapse supernovae; studied the decays of very proton-rich and neutron-rich light nuclei; and established a system for the production of rare isotopes that is needed to design experiments involving these isotopes.

THE K1200 ERA—1988–99

Early in this era, the K1200 fed beams to a limited Phase I.5 experimental area while the Phase II areas were prepared. These new areas held experimental devices that were much more powerful than those common in the K500 era: the A1200 fragment separator, the S800 spectrograph (later), and two large solid-angle detectors, the 4π Array and the Miniball. Together with the K1200 they made possible a much more varied and influential experimental program. A qualitative change, made possible by the A1200, was the explosion of research with beams of rare isotopes produced by bombarding a thick target, usually beryllium, with high-intensity beams from the K1200. The resulting array of fast-moving products, often more than 100 isotopes, was sorted by the A1200 to form a beam used for further studies. A careful choice of detection methods made possible experiments with extremely low intensities, often less than one particle/sec, and sometimes less than one/hr.

- HALO NUCLEI, nuclei with weakly bound neutrons, have extended distributions of almost pure neutron matter. Beams of ^{11}Li, ^{11}Be, and ^{19}C were produced and were studied as prototypes; it was shown that when their neutrons were removed, their momentum along the beam direction accurately reflected the momentum that these neutrons had in the nucleus. This finding was exploited by studying the breakup of many nuclei and using momentum distributions to determine orbital occupations. These results tested shell model descriptions of radioactive nuclei as transfer reactions had done for stable nuclei; it became possible to study nuclei far from the valley of stability that had unusual ratios of neutron number to proton number.

- EXCITATION OF COLLECTIVE MOTION IN NUCLEI by the electric fields of heavy targets (Coulomb excitation) was studied using 50–100 MeV/nucleon beams of radioactive nuclei. It had long been thought that nuclei with a magic number (2, 8, 20, 28, 50, 82, 126) of neutrons or protons were especially stable or strongly bound. This new and powerful technique hinted that the location of magic numbers depended on N/Z ratio in a nucleus; the venerable shell model had to be modified. Comparisons of proton scattering showed whether protons or neutrons contributed the most to excitation of low-lying states.

- TRANSPORT THEORIES (BUU theories) that describe the evolution with time of collisions of heavy ions were tested using the new 4π Array, the Miniball, and other specialized multi-element detectors: measurements of two protons emitted simultaneously elucidated the space-time evolution of particle-emitting volumes, and other measurements determined the energy at which nuclear attraction and repulsion balance. Both agreed in detail with the BUU theory developed at MSU. We now had a strong basis for understanding these reactions and using them to obtain other information. The rapid time scale of multi-particle breakup (multi-fragmentation) was established; nuclear thermometers were further refined; and differences in neutron-neutron and neutron-proton interactions in dense material were obtained; these differences are important for the evolution of neutron stars.

- NUCLEI IN THE COSMOS. The availability of radioactive beams opened a broad field of study related to stellar evolution, simply because the great majority of stellar reactions are among radioactive nuclei only transiently found in nature. ^{8}B was shown to have a proton halo, the first known, and its coulomb field–induced breakup was used to determine the production rate of neutrinos that were made in the sun and observed on Earth in neutrino detectors. The half-life of ^{44}Ti was used to study its production in supernovae, and we determined the contribution of $\alpha + \alpha$ collisions to the production of lithium in cosmic rays, and the masses of nuclei important in novae. The first high-resolution measurements of the weak interaction strength that describes the capture of electrons during supernovae core collapse were performed using the (triton, ^{3}Helium) reaction.

THE COUPLED CYCLOTRON ERA—2001–PRESENT

Coupling of the K500 and K1200, with a powerful ECR ion source and the A1900 fragment separator, led to increases in intensity of primary beams, which, depending on energy and particle, were ten or more times larger than those in the K1200 era; intensities of secondary beams were 1,000–10,000 times larger. These intensities and the rapid construction of this facility made the NSCL competitive and often dominant on the world scene. A secondary effect was that all but a few percent of experiments used secondary rare isotope beams rather than the primary beams from the accelerators. There was also an increasing emphasis on building highly efficient and specialized detection equipment laid out, as shown in figure 87, to permit experiments with the small intensities that could be obtained for beams of nuclei with extreme neutron-to-proton ratios. Many formerly impossible experiments became possible, resulting in studies of an exceptionally broad array of phenomena. An indication of the variety of rare isotopes available is shown in figures 88 and 89. In 2012–13, GRETINA, a state-of-the-art gamma ray detector built (mostly) at LBNL for a cost of about $20M, was at NSCL for a year and was used in twenty-three successful experiments. It is impossible to summarize these and other experiments in any efficient way, so I revert to a listing of topics to give a sense of the program's breadth and energy.

- EXTENT OF THE NUCLEAR SPECIES. For a set of isotopes with a given Z, the largest number of neutrons that can be bound defines the neutron drip line—added neutrons just "drip" away. A similar statement defines the proton drip line. Determining the extent of the nuclear species was a major focus, trying to answer the question: what nuclei lie within the neutron and proton drip lines? In 2007, Michael Thoennessen initiated a project to document isotope discoveries and publish discovery details.[190]

- EVOLUTION OF NUCLEAR STRUCTURE. A related question is how nuclear properties, shell structure and collectivity, change as one approaches the drip lines? What are the properties of neutron-unstable nuclei just outside the neutron and proton drip lines? How does the number of nucleons in different orbits vary with the N/Z ratio? How does the spin-isospin response of nuclei vary with and across nuclear shells? Long strings of isotopes are available for study using the CCF. For tin (Z = 50), for example, isotopes with masses from 99 to 137 and N/Z ratios from 0.98 to 1.74 are available, and permit the delineation of many properties that depend on this ratio. Such studies have greatly strengthened the earlier

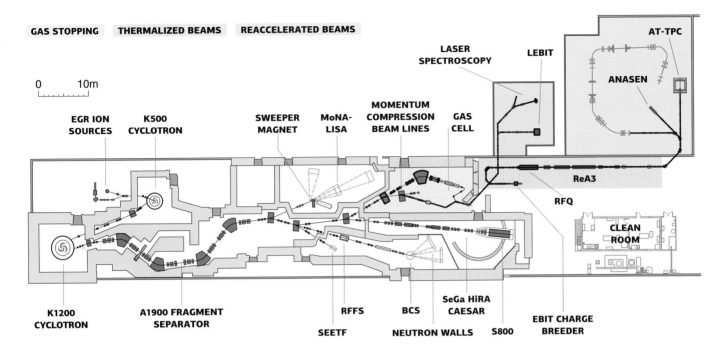

GAS STOPPING   THERMALIZED BEAMS   REACCELERATED BEAMS

FIGURE 87 (*above*). The layout of experimental equipment at the CCF. The cyclotrons and the A1900 fragment separator at the left deliver beams, mostly rare isotopes, to a large array of experimental devices. A list of these instruments and their applications is given in the later section: "The Present Coupled Cyclotron Era" within "Detection Apparatus Generations," on page 237.

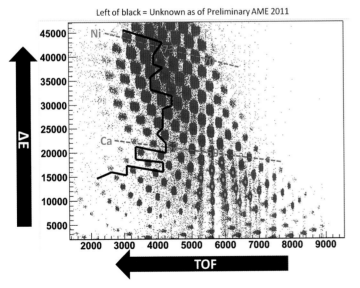

Left of black = Unknown as of Preliminary AME 2011

FIGURE 88 (*left*). This shows the variety of rare isotopes produced when one bombards a thick Be target with Se ions from the CCF. Measurements of the isotopes' velocity (time of flight) and the energy it loses (ΔE) in a thin detector labels different isotopes. It is a useful exercise to count the dots, each of which corresponds to a given isotope. For example, at least eight isotopes of calcium are produced in significant quantities. Often many isotopes can be studied at once. For example, when this experiment was done, all the rare isotopes to the left of the black line had unknown masses, which were then measured.

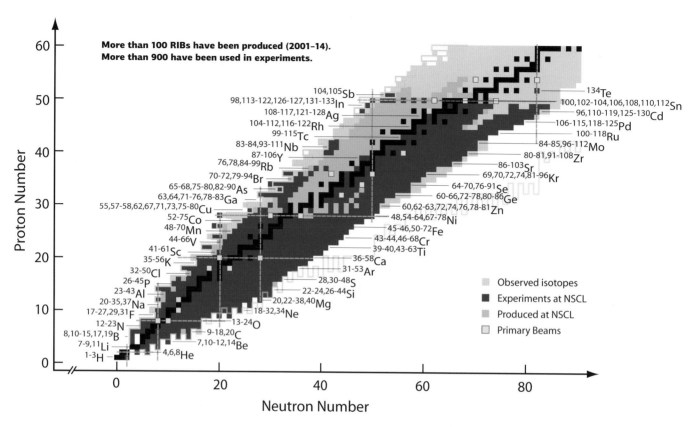

More than 100 RIBs have been produced (2001–14).
More than 900 have been used in experiments.

- Observed isotopes
- Experiments at NSCL
- Produced at NSCL
- Primary Beams

FIGURE 89. Production of rare isotopes in the CCF and FRIB.

Above: Isotopes produced, and/or used in experiments at NSCL. The primary beams accelerated by the cyclotrons are shown in yellow.

Right: The promise of FRIB. The area in dark blue shows the isotopes for which there is, at least, minimal information about nuclear properties. The isotopes in green, produced by FRIB, will more than double this number of nuclei, and its increased intensity will permit much more to be learned about those already known. FRIB will provide information about the great majority of nuclei with Z < 75 and half of even the heaviest isotopes.

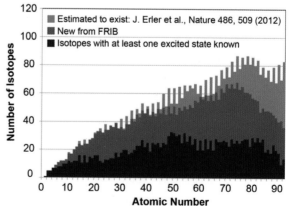

challenge to the shell model and made it clear that the magic numbers of the shell model depend on the N to Z ratio.

- NUCLEUS-NUCLEUS REACTIONS. Availability of long strings of isotopes and isotones made it possible to study the ratios of neutron and proton yields in reactions with projectiles having different N/Z ratios. These experiments determined the symmetry potential, which governs the radii and masses of neutron stars. Searches for the hadronic to quark-gluon phase transition continued at Brookhaven using RHIC.

- NUCLEI IN THE COSMOS. Nuclear processes are responsible for many phenomena in the cosmos, especially those involving explosions. For example, nuclear lifetimes are needed to accurately describe both the rapid neutron capture process that makes half of the elements heavier than iron and the evolution of light emission from X-ray bursters. Values of nuclear masses, of weak interaction strengths, and of nuclear reaction rates determine how the crusts of accreting neutron stars evolve and how elements, including of C, N, and O, the nuclei of life, are synthesized in novae and supernovae. The availability of nuclei far from the valley of stability has greatly increased the NSCL's experimental reach for studying such processes.

CROSS-DISCIPLINARY RESEARCH[191]

NSCL beams and facilities are sometimes used for non-nuclear research. Beam time requests for such experiments are handled in the same manner as other requests. They use a small percentage of the total beam time and often involve collaborations of NSCL and researchers in other MSU Departments or in external laboratories. Some examples:

During the K50 era there were extensive and pioneering studies of the metabolism of nitrogen fixation by blue green algae done in collaboration with scientists from the MSU-DOE Plant Research Laboratory. Later, nitrogen-13 labeled nitrogen nitrate produced using the K500 was used by scientists in the MSU Department of Plant, Soil, and Microbial Sciences to study the reactions that convert nitrogen fertilizer to gaseous nitrogen in the soil (see Chapter 7).

Radiation effects of heavy ions have been studied for beams incident on carbon nanostructures such as carbon nanotubes and onions,[192] on gallium arsenide wires,[193] and on DNA.[194, 195] The MSU College of Engineering and ANL have studied different aspects of the effects of defects induced in

A more detailed view, aimed at the expert, of the research activities involving the CCF and those doing it. The strong overlap of experimental and theoretical interests is evident. In the list below, experimentalists are denoted in roman type and *theorists in italics*. For full names and appointment periods, see figures 91 and 92.

- **How does subatomic matter organize itself and what phenomena emerge?**

  AT AND BEYOND THE DRIPLINES: Bazin, Gade, Kohley, Morrissey, Sherrill, Spyrou, Stolz, Thoennessen; *Brown, Hjorth-Jensen, Nazarewicz, Nunes, Zelevinsky*

  SHELL EVOLUTION, LEVEL SCHEMES, COLLECTIVITY: Bazin, Gade, Iwasaki, Liddick; *Brown, Bogner, Hjorth-Jensen, Nazarewicz, Zelevinsky*

  NUCLEAR WAVE FUNCTIONS THROUGH DIRECT REACTIONS: Bazin, Gade, Lynch, Tsang, Zegers; *Brown, Nunes*

  REACTION DYNAMICS, NUCLEAR EQUATION OF STATE: Kohley, Lynch, Mittig, Tsang, Westfall; *Danielewicz, Pratt*

- **How did visible matter come into being and how does it evolve?**

  COMPUTATIONAL ASTROPHYSICS, GALACTIC EVOLUTION, CHEMICAL EVOLUTION: *O'Shea*

  ORIGIN OF THE ELEMENTS: Hager, Schatz, Spyrou, Wrede, Liddick; *Brown, Nunes, O'Shea*

  NOVAE, SUPERNOVAE, X-RAY BURSTS: Hager, Schatz, Sherrill, Spyrou, Wrede, Zegers; *E. Brown, Brown, Nunes*

  NEUTRON STARS AND NUCLEAR EQUATION OF STATE: Lynch, Tsang, Schatz; *Brown, E. Brown, Bogner, Danielewicz, Hjorth-Jensen, Nazarewicz*

- **Are the fundamental interactions that are basic to the structure of matter fully understood?**

  CVC HYPOTHESIS, NEW INTERACTIONS AND COUPLINGS, EDM: Naviliat-Cuncic, Singh, Wrede; *Nazarewicz, Zelevinsky*

  DOUBLE BETA DECAY: *Brown, Bogner, Hjorth-Jensen, Nazarewicz*

  MASS MEASUREMENTS, IMME: Bollen, Wrede

  PRECISE MEASUREMENTS OF NUCLEAR RADII, MOMENTS: Mantica, Minamisono

- **How can the knowledge and technological progress from nuclear physics be used to benefit society?**

  MATERIALS UNDER EXTREME CONDITIONS: Bollen, Mittig, Stolz

  DETECTORS AND DATA FOR NATIONAL SECURITY: Gade, Iwasaki, Liddick, Sherrill, Thoennessen

  ISOTOPE HARVESTING: Bollen, Morrissey, Sherrill, Zegers

  REACTIONS FOR STOCKPILE STEWARDSHIP: *Nunes*

  FISSION: *Nazarewicz*

- **Accelerator Physics**

  ACCELERATOR PHYSICS AND PROJECTS: Wei, Yamazaki

  SUPERCONDUCTING RF: Facco, Saito

  BEAM DYNAMICS: Lund, Syphers

  HIGH PERFORMANCE ECR ION SOURCES

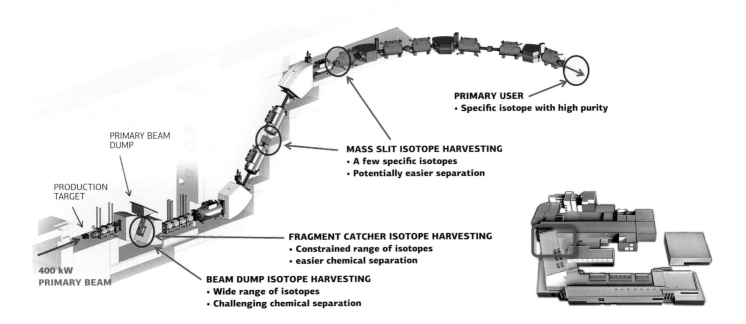

PRIMARY USER
• Specific isotope with high purity

PRIMARY BEAM DUMP

MASS SLIT ISOTOPE HARVESTING
• A few specific isotopes
• Potentially easier separation

PRODUCTION TARGET

FRAGMENT CATCHER ISOTOPE HARVESTING
• Constrained range of isotopes
• easier chemical separation

400 kW PRIMARY BEAM

BEAM DUMP ISOTOPE HARVESTING
• Wide range of isotopes
• Challenging chemical separation

superconducting materials.[196, 197] These studies make use of the extremely large localized deposits of energy and radiation damage along the microscopic paths of heavy ions passing through matter.

The laboratory also supports research in the national interest, but not subject to the PAC process, that is carried out on a cost-reimbursement basis. The study of Single Event Effects (SEE) in digital electronics, using irradiation by heavy ions to simulate effects seen in space, is the most common of these studies. A dedicated facility for this purpose was constructed by NASA and is used in collaboration with NSCL by Goddard Space Flight Center, Honeywell, and the University of New Hampshire's Space Science Center. Laboratory facilities were also used to test and calibrate detectors used in space flight and radiation monitoring applications.

Some of these heavy-ion irradiations will benefit FRIB and other high-power accelerators. For example, a terbium gallium garnet crystal was irradiated to test its suitability for use as a magnetic field sensor in high-radiation environments.[198]

FIGURE 90. A schematic view of the FRIB fragment separator located at the blue square shown on the picture of the FRIB complex. Its primary use will be to deliver beams to the primary user of the facility for research with rare isotope beams. However, and simultaneously, other isotopes can be harvested for use in medicine and other applications.

Studies of nuclear astrophysics would also fit in this category but have become a part of the mainstream laboratory program.

APPLIED ISOTOPE SCIENCE AT NSCL AND FRIB

Accelerators used by nuclear physicists have typically served one experiment at a time. At NSCL and eventually FRIB, there is a major effort led by Georg Bollen to make better use of the available beams and the investments in their construction and operation, by serving more than one simultaneous purpose. This might include providing beams to more than one nuclear science user, simultaneously harvesting isotopes for use in medicine and other applications, and studying the material science properties of matter under the extreme conditions that are produced when materials are bombarded by heavy ions. During operation of the NSCL facility for delivering fast beams or stopped beams for science, the ReA3 reaccelerator can be available for other purposes.

Figure 90 shows how FRIB has been designed with simultaneous (commensal) use in mind. Present concepts include producing isotopes in the water-filled primary beam dump where the beam is stopped in the FRIB fragment separator. One can also use isotopes caught on mass-defining slits by a variety of on-line devices. For example, a helium jet system could transport collected isotopes to an ion source where they would be used to produce very low-energy beams for stopped beam experiments or reacceleration. When not being used for the acceleration of rare isotopes, the ReA3 reaccelerator facility can produce energetic stable heavy-ion beams for the study of matter in extreme conditions and other material science and engineering applications.

- ISOTOPE HARVESTING. The FRIB design has provisions that will allow isotope harvesting in the future, while the NSCL provides an opportunity for early tests of some of these concepts. DOE has funded an R&D project on isotope harvesting from water, a Washington University, Hope College, and MSU collaboration.[199] This project includes studies of how a particular isotope can be extracted from the water target in a useful form.
- EXTREME MATERIALS RESEARCH. There is great interest, at MSU and elsewhere, on studying the behavior of materials when bombarded by intense beams of heavy ions. NSCL is one of the few facilities worldwide that can provide these beams over a very large energy range, including low-energy beams from NSCL's ReA3 reaccelerator. MSU has unique facilities and broad expertise in material science and engineering, and plans to add up to ten faculty

positions involving other MSU Departments to support cross-disciplinary research with radioisotopes and heavy beams; other universities are also likely to add positions. Some research is already underway at the NSCL through the interdepartmental MATX Program, which is funded by a Strategic Partnership Grant from the MSU Foundation. The research includes studies of ways to improve graphite's lifetime in high-radiation environments, the resistance to high-cycle fatigue and radiation of titanium alloys, and the performance and lifetime of diamond detectors in high-radiation environments.

Looking Forward: What's in It for Us—
The Nation and Society?

The research programs just described—and indeed this entire book—for many will seem to be a story of complex research with complex instruments yielding esoteric results of interest mostly to nuclear scientists. But it is a real story, one that will drive the new discoveries and yield the new knowledge that are the promise of FRIB to researchers.

One might compare the search for the nature of matter to the search for an understanding of history, or of the human mind, or of the origin of *Homo sapiens*. It is a topic of intrinsic interest to the curious human species.

We want to know, for example, how the elements we see about us were formed, the true extent of the nuclear species as estimated in figure 25, and why it is that rare isotopes behave in the subtle ways we observe. Indeed, we have pushed the boundaries of the nuclear species far beyond what we knew ten years ago, and FRIB will permit us to observe the great majority of all species predicted to exist by our present models of nuclei.

We have already shown that a long-standing belief that magic numbers are universal is false. They are useful approximations for stable nuclei, but they may offer little predictive power for rare isotopes. FRIB will provide the tests of our nucleic descriptions with much greater precision and sensitivity, and this will open new frontiers of knowledge and understanding of the universe.

Another major contribution of FRIB to society will be the education of highly competent people whose skills turn out to be useful in many technical venues. Some of them return to universities or national laboratories where they educate still another generation of knowledge-seeking students. Others will devote their careers to activities crucial to national goals.

The DOE's mission is "to ensure America's security and prosperity by addressing its energy, environmental, and nuclear challenges through transformative science and technology solutions."[200] The capability of nuclear scientists is well tuned to meeting this goal through developing tools and knowledge relevant to energy sources, manufacturing, medicine, and national security. Better nuclear models will allow optimization of the next generation of nuclear reactors and the evaluation of techniques for destroying nuclear waste. Beams from the CCF and FRIB will allow us to probe advanced materials, some of nanoscale size, to determine how they are affected by radiation.

Another example, familiar to many of us, is the multi-billion-dollar industry devoted to radiopharmaceuticals—drugs containing radioactive isotopes—that are used in a great variety of diagnostic and therapeutic procedures. FRIB will generate research samples of many new radioisotopes for evaluation of new approaches to therapy or diagnosis.

So what's in it for us in this fifty-year advance of the frontiers of nuclear science? And in building "up from nothing" a series of MSU cyclotrons, and now the Facility for Rare Isotope Beams? At the very least, contributions to knowledge and culture, the education of talented individuals, and the development of technologies that can impact our lives directly in many ways in the decades and centuries ahead. What lies still beyond that will become clear in far future years. That is the nature of fundamental research.

SUPPLEMENT

Behind-the-Scenes Views of the Cyclotron Laboratory

■ **NSCL Generations**

The evolution of the Cyclotron Laboratory is reflected in the evolution of its staff and scientific facilities. This section comprises brief descriptions of their changes as the laboratory grew and its research focus changed.

PEOPLE GENERATIONS—FACULTY AND SENIOR RESEARCHERS

Talented people are the drivers of all else that happens. This truism drove the laboratory, as it grew, to devote an unusual amount of attention to maintaining an effective mix of the young and the experienced. This required the development of new faculty systems as we describe below.

Most new laboratories are led by a relatively senior individual who then hires a group of researchers to build the new facility. The situation at MSU was different. It was a fledging research university, with a physics department just beginning to develop an emphasis on research, and the Laboratory Director was himself only a few years past graduate school. He hired a small group of accelerator physicists who arrived between 1958 and 1961. Soon thereafter the laboratory attracted a group of young nuclear physicists who were willing to chance coming to a new facility and who saw the opportunity of building a laboratory from scratch. They were of similar ages, around thirty, and many had little experience with using cyclotrons. The cyclotron itself was uniquely capable, and this first generation of experimental faculty became closely knit and dedicated to performing research of high quality. The reputation of the laboratory grew rapidly and made it possible to compete for the new facilities that have been described. Many of this first generation remained at MSU until retirement and played important roles in the first three of the accelerator generations described below. Of course, their interests and capabilities had to evolve, and did so effectively.

It slowly became clear, however, especially as the end of the K50 era approached, that this initial group did not provide a path to the long-term future. An ideal age structure in a forefront scientific laboratory has a combination of the wisdom of the experienced and the new abilities, enthusiasm, and idealism of the young. The lab was experienced, but lacked a young contingent; there had been no new hires for seven years, since 1970, nor were tenure lines available. To address this problem, an NSCL-unique (at MSU) Continuing Appointment system was formed that allowed the laboratory to recruit a group of exceptional young researchers and to nurture them until positions in the Physics and Chemistry faculties opened up with departures and expansion. These individuals and the refilling of open positions carried us through the third (K1200) phase, to 2000, by which time nearly all of the first generation had, or was about to, retire and the second generation had become a senior generation.

Again it was apparent that a large group of new people was needed to carry the laboratory into its next phase, especially as many of the second generation had moved into the administrative ranks necessary to compete for the major facility that has become FRIB. We also had to attract accelerator physicists with the competence to design, fund, and build FRIB and to build the world-competitive research program in accelerator physics that is important at a major accelerator facility. During the first phases of the FRIB project, when funding was uncertain, this was particularly difficult. Offers of joint appointments in the NSCL and an academic department were necessary to make these positions sufficiently attractive. A new NSCL Faculty system, with rolling tenure, described in Appointment Systems, was created to remedy this situation and now includes twenty positions. It has allowed the NSCL to attract the best of the new generation of nuclear and accelerator researchers.

Figures 91 and 92 show the tenures at MSU of NSCL researchers who held the position of Professor, or other influential senior positions in one of the three academic categories. Dates are taken from the years of appointment and departure; having been taken from a variety of sources they may not always be accurate. Hence the results are meant to show patterns rather than a record for individuals.

The NSCL as a whole has grown greatly since its early days. In 1963 (1970), about thirty-five (sixty) people worked at the NSCL; as of June 2013, that number was about 420 (plus ninety undergraduate students) and growing. It includes forty-one faculty, mostly Physics and Chemistry Department Faculty, eighty-eight appointments in the CA system, fifty-seven graduate students, twenty-eight Research Associates, 158 support staff in MSU-wide employment systems (Administrative-Professionals, Clerical-Technical, and Skilled Trades), and forty-five others: scientific visitors, consultants, and MSU personnel connected with the NSCL or FRIB.

PEOPLE GENERATIONS—STUDENTS AND POSTDOCTORAL ASSOCIATES

Graduate Students:[201] How Their Experiences Have Evolved

When the laboratory began operation in the early 1960s, most research in nuclear science comprised studies of nuclear structure. Although the laboratory was new, it had a superior facility for structure studies and attracted graduate students of high quality. The general expectation was that students would design and build any special equipment they needed for their experiments and independently take and analyze their data and publish it, usually with a small number of co-authors. When the first wave of Cyclotron Laboratory PhDs was awarded around 1970, U.S. PhD production was high and academic jobs were relatively scarce. However, the balance of technical and nuclear science training our graduate students had received served them well and led to successful careers, especially in national laboratories and industry.

The graduate student experience changed as the laboratory evolved into the user-facility era; experimental collaborations became larger on average, the cost of an hour of cyclotron time grew, and the award of time depended on decisions of a Program Advisory Committee (PAC). Nevertheless, most students had a similar experience, although often in group collaborations where tasks were more regularly shared. Apparatus building was still common, however, as was responsibility for beam-line operation.

By the 1990s, experiments with the fragment separators—the A1200 and A1900—dominated the program and these responsibilities evolved further. Beam preparation became a job performed predominately by NSCL professionals and data-taking was a team effort. There were still many opportunities for apparatus development, but the balance of graduate student effort shifted toward

FIGURE 91. Time at MSU of theoretical faculty. The ends of the bars show the beginning and end of their tenure at the cyclotron laboratory and the numbers at the far right, the length of this tenure.

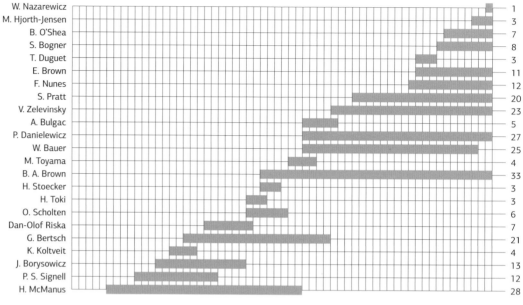

FIGURE 92. Time at MSU of experimental and accelerator faculty and senior researchers.

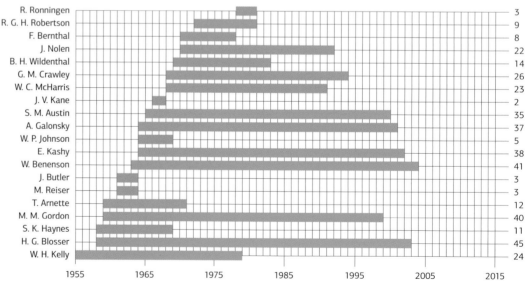

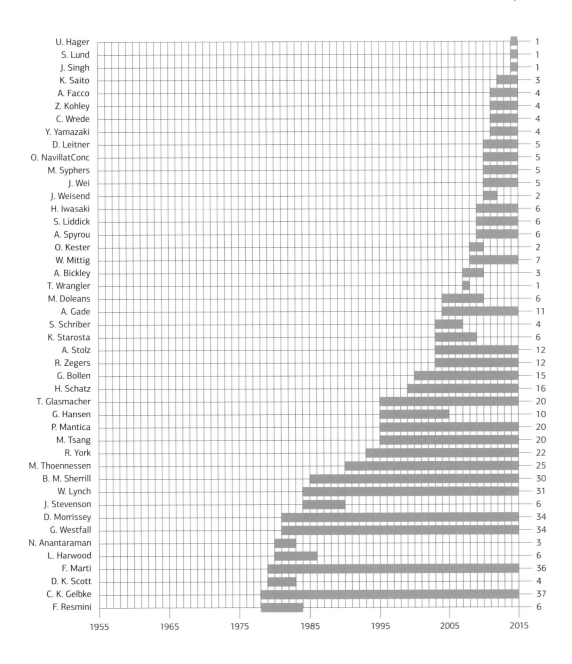

experiment planning, apparatus simulations, and analysis of complicated multi-dimensional data sets. Since these changes reflected those in the field as a whole, their graduate-student experience prepared students well for post-PhD life.

There were other related changes that had to be faced. One was the increasing time spent in obtaining a PhD, nationally an average of almost seven years. A major effort was made to shorten this time by laying out expectations more carefully (in a graduate student handbook, for example), restructuring course and examination schedules, tracking progress toward degree on a regular basis through meetings and progress reports, and avoiding delays in scheduling thesis experiments where possible. For the 1998 to 2002 period, the average time from entering graduate school to obtaining to a PhD at the NSCL was reduced to 5.7 years, and for 2008 to 2014, it further decreased to 5.3 years—both significantly shorter than the seven-year national average.

Another issue was to ensure that students' overall experience would prepare them for a scientific career in which both individual initiative and ability to work in groups were important. A set of expectations was formulated and set out in the NSCL's Graduate Studies—2013:

> Graduate students in experimental nuclear science will be involved in all aspects of performing an experiment at NSCL: writing a proposal which will be reviewed by the external Program Advisory Committee, designing an experiment, setting up the hardware and electronics, writing data acquisition and analysis code, analyzing and interpreting the results, and finally writing a manuscript for a peer-reviewed journal.
>
> Theory students have access to world experts who frequently visit the laboratory and collaborate closely with local faculty.

Students participate in all experiments of their group and are encouraged to broaden their view of nuclear science and its techniques by participating in experiments of other groups. Most students do so, especially early in their graduate career. Students are represented on most laboratory committees; administer their own office space, furniture, and desktop computers; organize their own seminar series; participate in outreach activities inside and outside the NSCL; and have weekly meetings. Students are sent to conferences to present their work and, when appropriate, to visit other institutions to further their research interests.

That our students are in high demand, generally having several job offers at places they want to work, seems to justify this approach. Snapshots of the post-PhD careers of Cyclotron Laboratory PhDs are shown in figure 86; the fraction working in industry has decreased in recent years and the number working at national laboratories has greatly increased.

Although the number of graduate students working in the various areas of nuclear science nationwide has changed little over the past thirty years, the number of graduate students at MSU has grown, reflecting the greater research capability and visibility of the NSCL. In 1970, 1980, and 1993, there were 26, 28, and 37 NSCL graduate students, respectively, at the NSCL and the production of PhDs was about half the present level. The laboratory now has about sixty graduate students: 70 percent in experimental nuclear or nuclear astrophysics, 20 percent in theoretical nuclear physics or nuclear astrophysics, and 10 percent in accelerator physics. Most students (75 percent) are U.S. citizens. About half are supported by the main NSCL grant from NSF, others by JINA-CEE, MSU, NNSA, and DOE funds.

Undergraduate Students

Although the laboratory has always employed undergraduates in a variety of roles, as the laboratory has grown this group has become surprisingly important and large. At present around ninety undergraduates are employed at the laboratory during the academic year and somewhat fewer during the summer. They come from a variety of majors; a snapshot listed employees with majors in

Accounting, Advertising, Astrophysics, Biochemistry and Molecular Biology, Chemical Engineering, Chemistry, Civil Engineering, Computer Engineering, Communicative Sciences and Disorders, Construction Management, Childhood Development, Dietetics, Electrical Engineering, Family Community Services, Hospital Business, Human Biology, Human Resources and Labor Relations, Kinesiology, Mechanical Engineering, Media Arts, Media Technology, Physics, Philosophy, and Psychology.

Students work in a variety of roles, some with experimentalists or theorists, some in the technical support functions of the laboratory, and some in administrative functions.

There are also more formal programs. Some ten undergraduate students spend their summers at the NSCL in a REU (Research Experience for Undergraduates) program sponsored by NSF, NSCL, and the MSU Department of Physics; they have the experience of carrying out and publicizing, through posters and presentations, a small experimental or theoretical project. A few students also write senior theses, or serve as Professorial Assistants to faculty members, in a program for high-achieving students sponsored by MSU.

Postdoctoral Researchers[202]

The nature of the experiences of a postdoctoral researcher and a graduate student at the NSCL are somewhat parallel. The exception is that the postdoctoral experience is viewed as the last step in an individual's preparation to be an independent researcher. Postdocs are given substantial independence, are expected to work with much less oversight than a student, and to play a role in the supervision of graduate students during experiments and the analysis of data. They are often given responsibilities for construction or oversight of major pieces of experimental apparatus and are expected to present their research at conferences and to be able to defend the work of their groups.

A typical postdoctoral experience lasts three years. Around 60 percent of our postdocs move on to coveted positions in academic institutions, as shown in figure 86,[203] reflecting their abilities and the quality of their NSCL experience.

The number of postdoctoral researchers at the NSCL has increased greatly. It was about six, seven, and ten, in 1970, 1982, and 1993, respectively. At present, the laboratory has about thirty postdoctoral researchers, 60 percent in experimental nuclear or nuclear astrophysics, 25 percent in theoretical nuclear physics or nuclear astrophysics, and 15 percent in accelerator physics.

THE UNDERREPRESENTED

When the Cyclotron Laboratory was founded, almost no women or minorities worked at the laboratory. Since then the situation has changed in a positive direction but is clearly a work in progress.

Most promising for the future of a more representative workforce is that the fraction of female graduate students in the last five years has varied between 20 and 35 percent, and presently is 30 percent. The fraction of female research associates has been highly variable because of the small numbers involved, ranging from 6 percent to over 35 percent. There are now five female faculty members, three holding leadership positions in the laboratory: Chief Scientist, Associate Director for Education, and Theoretical Nuclear Science Head. The fraction of minorities has varied from 15 to 20 percent, mainly of Asian descent.

MSU "GRADUATES" IN SENIOR SCIENTIFIC OR ADMINISTRATIVE POSITIONS

Although not usually thought of as part of the educational role of the Cyclotron Laboratory, a rather large number of those who contributed to the research program at the Laboratory have taken important positions elsewhere. This is a comment on both the quality of people who choose to spend time here and on the value of their MSU experience. With no attempt to be inclusive, some individuals are listed below.

Frederick M. Bernthal	Faculty	Chief Leg. Asst. for U.S. Senate Majority Leader; Member Nucl. Reg. Comm.; Asst. Sec'y State; Dep. Dir. NSF; Pres. Univ. Res. Assoc.
George F. Bertsch	Faculty	Fellow INT; Bonner Prize
Gerard M. Crawley	Faculty	Dean Grad Sch. MSU; Dean, Coll. Sci. Math., Univ. South Carolina; Acting Director, Physics Div. of NSF; Head Frontiers Engineering Directorate, Sci. Found. Ireland
Robert Doering	Student	Texas Instruments Senior Fellow; Chair TI Technical Board; U.S. Member Int'l Roadmap Comm.
Sydney Gales	Postdoc	Dir. CNRS; Dir. GANIL; Dir. IPN; Dep. Dir. IN2P3; Chair of NuPECC
Barbara Jacak	Grad Student	Member NAS, Dir. Nuclear Physics Div. of LBNL
Oliver Kester	Faculty	Head of the FAIR Project at GSI
Lawrence Kull	Grad Student	Founder and COO, Science Applications Int'l Corp.
Jerry A. Nolen	Faculty	Director of ATLAS at ANL
Dan-Olof Riska	Faculty	Director Helsinki Inst. Physics
Bruce Remington	Student	Chair APS Topical Group, Plasma Physics; 2011 Edward Teller Medal
R. G. H. Robertson	Faculty	Director CENPA, Univ. Wash.; Chair NSAC; member NAS; Bonner Prize
David K. Scott	Faculty	MSU Provost; Chancellor U. Mass.
Horst Stoecker	Faculty	Director GSI; V.P. Helmholtz Assoc.
Hiroshi Toki	Faculty	Director RCNP (2001), Osaka Univ.

ACCELERATOR GENERATIONS

The construction of the K50 cyclotron was driven less by science than by the ambition of President John Hannah, the MSU administration, and the Physics Department to build a credible research institution. Henry Blosser, while interested in the science, found most satisfaction in building powerful new accelerators. As time went on, the direction of accelerator development was driven more and more by the science interests of the faculty.

We have described the several generations of accelerators that have been built at the NSCL: the K50, K500, ECR + K1200, and K500 + K1200. By the time FRIB comes into operation, these devices will have operated for about fifty-three years in the fifty-seven-year period from 1965 to 2021 in four different generations. It took a total of around five years to install the new generations of accelerators and beam lines, during which a facility was inoperative.

GENERATION	OPERATED	YEARS	SPECIALTIES
K50	1965–79	15	LI, NS, NAP
K500	1982–90	8	LI, HI, NS, NM, NAP
K1200 + ECR	1988–90 (I.5) / 1990–99 (II)	2/9	HI, RB, DL, NM, NAP
K500 + K1200 + ECR	2001–20 (?)	19	RB, HI, DL, NS, NM, NAP
ReA3	2014–		RA, NAP, NS
FRIB	2022–		RB, RA, HI, DL, NS, NM, NAP

LI = Light Ions: Beams of neutrons, protons, deuterons, alpha particles; NS = Nuclear structure: Studies of the quantum states of nuclei; NAP = Studies of nuclear reactions that occur in the cosmos; HI = Heavy Ions: Beams of heavy nuclei, greater intensity in later accelerator generations; NM = Nuclear Matter: Study of the gross properties of heated nuclear material; RB = Radioactive, Rare Beams: Production of beams of unstable nuclei; DL = Drip line nuclei, study of nuclei at the limits of stability. Growing greatly in importance in later accelerator generations; RA = Reaccelerated lower energy beams.

ION-SOURCE GENERATIONS, ESPECIALLY OF ELECTRON CYCLOTRON RESONANCE (ECR) SOURCES

ECR ion sources played a large role in the evolution of the Cyclotron Laboratory. During the K50 era and the first part of the K500 era, the laboratory used standard Penning (PIG, Phillips Ion Gauge) sources placed in the magnetic field at the center of the cyclotron. They served well for the production of ions from light gaseous elements, but for heavy ions they had short lifetimes and limited intensities.

The invention[204] of ECR sources relieved these limits, providing lifetimes of weeks rather than hours and much higher intensities of highly charged heavy ions. The specialized magnets that confined the gaseous source plasma had to be placed outside the cyclotron, as did the external source of microwaves that ionized the plasma. Beam lines led beams extracted from the ECR into the cyclotron along its symmetry axis.

When the NSCL built its first ECR source in the mid-1980s, these sources were at an early stage of development and the dependence of output current on source properties was not established. The prevailing wisdom was that current increased with size and radio frequency. We decided to test the size-dependence by building an ECR source (called the RT-ECR) somewhat larger than the successful source then operating at LBL. Increasing the frequency was more difficult; high-frequency microwave sources are very expensive, and the required magnetic field strength increases with frequency, probably mandating superconducting magnets to confine the plasma. It was thought that a superconducting source would be the second step.

After operation of the RT-ECR began, we realized that long beam development times made it difficult to operate a research program with a single ECR. Fearing that it would take a long time to design and build a superconducting ECR, we opted instead to build a compact room temperature source (CP-ECR) specialized to the production of light ions, especially lithium; the superconducting source (SC-ECR) was delayed until 1990. Later two room temperature sources, ARTEMIS A and ARTEMIS B, and a second superconducting source, SuSI, were built. The two Artemis sources were built in collaboration with the LBL group; Artemis A and SuSI now provide beams for the coupled cyclotrons.

The desired capability of the ECR sources changed with time. For a single cyclotron system, such as ECR + K1200, high intensities of highly charged ions are needed to produce usable high-energy beams. For coupled systems, such as the ECR + K500 + K1200, the ion charge is increased upon injection into the second cyclotron when it passes through a thin stripping foil and loses electrons. One, therefore, needs instead very high intensities of ions with lower charge. For practical sources, intensities typically increase by a factor of ten or more, and one obtains much higher energies as well. For either approach, highly precise (low emittance) beams increase the efficiency of transport through the cyclotron(s) and the final beam intensity. High efficiency has the great benefit of reducing undesirable heating, which can melt internal cyclotron components. Much recent work on these sources has centered on means of producing metallic ions and on increasing beam quality as well as intensity.

We now know that output currents, at least for highly charged ions, increase with about the square of the source frequency, but the source size dependence is still not firmly established. A critical point is the amount of power one can couple into the plasma, about four kilowatts for SuSI at present, and eight kilowatts for the world-leading source, VENUS, at LBL. SuSI operation at twenty-four GHz is being tested. As of August 2014, argon and oxygen have been used and match or exceed results from VENUS at LBL.[205] SuSI might be adequate to meet the intensity specifications for FRIB, but other operational considerations—increased source size to accommodate a variety of ovens for different elements and the need for a different cryogenic supply—mean that a larger, more VENUS-like source will probably be built. The sources used to date are:

ACCELERATOR	SOURCE	FREQUENCY	DATES USED
K50	Penning		1965–79[a]
K500	Penning		1979–85[a]
	RT-ECR	6.4 GHz	1985–90
	CP-ECR	6.4 GHz	1987–90
ECR+ K1200	RT-ECR	6.4 GHz	1990–99
	CP-ECR	6.4 GHz	1990–95
	SC-ECR	6.4 GHz	1990–99[b]
ECR + K500 + K1200	SC-ECR	6.4 GHz	2000–08
	ARTEMIS A	14.5 GHz	2000–
	ARTEMIS B[d]	14.5 GHz	2005–
	SuSI	14.5, 18, 24[c] GHz	2009–

a. Initially the Penning sources were used for isotopes of hydrogen and helium. Beginning about 1977, they were also used to produce light heavy ions. Their use was terminated when RT-ECR became available in 1985.

b. The SC-ECR was originally designed to operate at fourteen GHz, but the superconducting hexapole coils failed to operate at the high fields required for this frequency.

c. Twenty-four GHz operation is being tested. As of August 2014, argon and oxygen have been used and match or exceed results from VENUS at LBL.

d. Artemis B is identical to ARTEMIS A and is mainly used offline for studies aimed at optimizing source performance and injection into the K500.

DETECTION APPARATUS GENERATIONS

Different accelerators producing different beams or energies and studying different phenomena require different experimental devices. Some of these will be used by a large number of researchers. The Enge spectrograph of the K50 era, the A1200 of the K1200 + ECR era, and the A1900 and S800 spectrograph of the CCF era are examples of these. Others may be specialized and serve one or a few users. The Penning trap, LEBIT, used for mass measurements during the CCF era, is an example of the latter. The growth of research with radioactive beams was a major driver for change; detectors had to become highly efficient to extract interesting science from the very weak beams. In the K50 era, most beams of interest had intensities of a billion (10^9) particles per second or more. In the coupled cyclotron era, many experiments have to be done with intensities of less than 10,000 per second and some information can be obtained with intensities of 1 per hour or less. Another reason for high efficiency is the growing cost of operating the larger accelerators. If a detector is twice as efficient, after debugging the experiment, data can be collected in half the time, greatly increasing productivity and reducing the cost of each experiment.

The K50 Era

In this era, most experiments were done with light-ion beams, typically protons or neutrons, beam intensities were large, and only one or a few products were produced in a nuclear reaction. A magnetic spectrometer like the Enge might sample only 1/10,000 of the space about the target, but its high resolution usually made it possible to identify the particle and its energy. Rates were high and even though many samples at different angles had to be taken, experiments could usually be done in a day or two.

Many experiments used a simple combination of two or more particle detectors, called a particle telescope, to identify product nuclei and obtain resolution sufficient to study individual nuclear states. These particle telescopes consisted of a thin detector backed by a thick detector; particles of interest passed through the thin detector, depositing an amount of energy that depended on their charge (their atomic number). They then lost the rest of their energy and stopped in the thick detector. Combining the thin detector signal (ΔE) and the thick detector signal allowed one to obtain the particles' energy and identity.

Gamma rays could also be detected with one or two detectors and yielded sufficient efficiency and resolution. Scattering chambers, basically evacuated cylinders, to house these types of devices were widely used. In all these experiments, to sample different angles one moved the detector. Neutrons were a special case: one measured their energy by measuring their velocity. To get an accurate result the detector had to be far from the target, sometimes as much as sixty feet away. In order to sample different angles one left the detector in place and rotated the beam using a question mark–shaped beam swinger. The suite of principal detectors included the Enge Split-Pole spectrometer, a forty-inch scattering chamber, a gamma-ray goniometer, a zero-degree neutron detector, and the beam swinger.

The K500 Era

Because the K500 cyclotron system was built on such a small budget, it was necessary to reuse much of the detection apparatus that was carried over from the K50 era and to build detectors using spare parts from other laboratories. Intensities were relatively large for most beams. However, the energies were large, and the Enge Split-Pole magnet was not always powerful enough to handle reaction products of interest. For this purpose, a spectrograph, the S320,[206] see figure 52, was constructed with magnets available, used and mostly free, from other accelerator laboratories. The S320 was used until 1990. A reaction products mass separator (RPMS)[207] was constructed, again mostly of spare parts, to provide samples of stopped radioactive nuclei of known mass for studies of their decays. As heavier beams were accelerated, an interest in studying the properties of nuclear matter grew, and this required the ability to determine the mass and charge of the many products of a single collision. Initially these studies were done with small arrays of detectors, usually less than fifteen, in a scattering chamber. Two devices sensitive over almost the full angular range in three dimensions (4π steradians): the 4π Array (built by Gary Westfall and his collaborators)[208] and the Miniball (built by Romualdo de Souza and his collaborators)[209] were developed during this era, but were mostly used later with the K1200.

The 4π Array was an ambitious device and could detect particles from protons to uranium, using a stack of detection elements with different properties. It was modular in nature, and because of this could be supplemented with other special purpose, or higher granularity detectors. It could also handle high event rates, and could be used to characterize the nature of collisions that led to improbable events. The 4π's first uses, before it was completely finished, were with the K500 and in Phase I.5. Later it was used in a variety of experiments that determined the nature of collisions of heavy ions and the properties of compressed and heated nuclear material.

The Miniball had properties somewhat complementary to the 4π Array. It also had large-solid-angle coverage, but its 188 detectors gave higher resolution and more capability to distinguish different products up to Z = 18. It could not, however, run at such high rates and was less able to accommodate other detectors. It was transportable and was used at GSI and in Saclay as well as at MSU. The Miniball was used in a variety of experiments and was sometimes placed inside the Superball (see later) to provide measurements of neutrons and charged particles simultaneously.

The K1200 + ECR Era (Including Phase I.5)

During the early part of this era operation was in a small vault containing the 4π Array and a large ninety-two-inch scattering chamber which could hold a variety of devices. When full operation began in October 1990, implementation of an impressive array of detectors began. The A1200 fragment separator was probably the most important new device. Other general purpose apparatus included the 4π Array (supplemented by special purpose devices), the Miniball, and continuations from earlier eras: the ninety-two-inch scattering chamber, the RPMS, and the S320 spectrometer. Also completed in this era, were a neutron-wall system, built by Aaron Galonsky and colleagues with position sensitivity and capability of distinguishing neutrons from gamma rays; the S800 spectrometer; and a Beta-Nuclear Magnetic Resonance apparatus to measure ground state magnetic moments built by Paul Mantica and his collaborators. Some major equipment was built by outside users: BIGSOL, a high magnetic field solenoid[210] served as a high-efficiency fragment collector for use with radioactive beams, and the Rochester Superball,[211] a segmented device that contained 17,000 liters of gadolinium-loaded scintillator and could detect the number (multiplicity) of neutrons formed in a reaction. It had an interior cavity large enough to hold the Miniball, thus allowing one to observe protons and neutrons simultaneously. The S320 was removed from service in October 1995; its function was taken over by the S800 spectrograph.

The Present Coupled Cyclotron Era

During this era the variety of detectors increased greatly and it would require too much space to describe them all, or to make their principle of operation easily available to the non-expert. We list them here to give a rough feeling for the nature and variety of detectors required in a forefront nuclear science laboratory. The evolving list of available instruments, together with detailed descriptions, is available on the NSCL website: www.nscl.MSU.edu/users/equipment.html. Cites to the builders and a description of the apparatus are also given there in most cases.

As of December 2014, the list is:

- FIFTY-THREE-INCH CHAMBER. General purpose vacuum vessel used for mounting large detector arrays surrounding a central target.
- A1900 FRAGMENT SEPARATOR. Large-solid-angle, high-magnetic-rigidity projectile fragment separator.
- GAMMA-RAY-ENERGY TRACKING ARRAY (GRETINA). A national resource that was used at NSCL for about twelve months during 2012–13. It was then moved to ANL and will return to MSU/NSCL for twelve months of operation beginning in mid-2015.
- SEGMENTED GERMANIUM ARRAY (SEGA). Eighteen thirty-two-fold-segmented high-purity germanium detectors for in-beam γ-ray spectroscopy with fast exotic beams.
- THE BETA-NUCLEAR MAGNETIC RESONANCE APPARATUS. Beta telescopes in a magnetic field to measure ground state magnetic moments.
- BETA COUNTING SYSTEM (BCS). Stack of silicon detectors to study the beta decay of radioactive species produced by fast fragmentation.
- HIGH RESOLUTION ARRAY DETECTOR (HIRA). Twenty telescopes, each consisting of two silicon-strip detectors backed by four CsI (Tl) crystals.
- MODULAR NEUTRON ARRAY AND LARGE MULTI-INSTITUTIONAL SCINTILLATOR ARRAY (MoNA-LISA). A pair of detector arrays with a total of 288 plastic scintillators for detection of high-energy neutrons with high efficiency. Built by a large collaboration of, mainly, undergraduate colleges.
- NEUTRON WALLS. Two large-area, position-sensitive neutron detectors with neutron-gamma selection.
- NEUTRON EMISSION RATIO OBSERVER (NERO). Detector for low-energy neutrons.
- S800 SPECTROGRAPH. High-resolution, large-acceptance spectrograph.
- LOW ENERGY BEAM AND ION TRAP (LEBIT). A device to slow rare isotope beams to low velocities. Combined with a Penning trap for high-accuracy mass measurements.
- THE SWEEPER MAGNET. Compact large-gap superconducting dipole magnet for use with MoNA-LISA.

- RF FRAGMENT SEPARATOR. Parallel-plate separator with radio frequency field to purify proton-rich rare isotope beams.
- BEAM COOLER AND LASER SPECTROSCOPY (BECOLA) END STATION. A facility for laser spectroscopy and beta-NMR studies.
- TRIPLE PLUNGER FOR EXOTIC BEAMS (TRIPLEX). A plunger for lifetime measurements with the recoil distance method (RDM).
- SUMMING NAI (SUN) DETECTOR. A gamma-ray total absorption spectrometer.
- LOW-ENERGY NEUTRON DETECTOR ARRAY (LENDA). An array of twenty-four scintillator bars for detecting 0.15–10 MeV neutrons.
- URSINUS COLLEGE LIQUID HYDROGEN TARGET. A target cell that maintains liquid hydrogen at about 18 K.
- SINGLE EVENT EFFECTS TEST FACILITY (SEETF). A dedicated in-air irradiation station with diagnostic equipment and controls.

■ Laboratory Infrastructure

Laboratory infrastructure is the foundation upon which its operation depends. The following is a brief description of the principal items.

COMPUTER SYSTEMS

The change in the nature of computation over the life of the laboratory has been remarkable. The following discussion is based partly on information provided by Ron Fox.

The earliest cyclotron design work was described earlier in this book, and involved the use of the MISTIC (Michigan State Integral Computer) computer, built at MSU in 1956–57.[212] It was ten feet high, eleven feet wide, and two feet thick, and was a state-of-the-art device. It had a memory of only a thousand words (later expanded to around 16,000 words), and multiplication took a millisecond. There were no direct input terminals, video displays, or magnetic tapes; input and output involved punched paper tape and, later, punched cards. Programming was directly in machine language—no Basic or FORTRAN or C++.

A major change occurred in 1963, when MSU purchased a Control Data Corporation 3600 computer with 32K memory and eight tape units. But all computing still took place at the remote computer lab site. It was an open question, in 1963, as to whether special access to the 3600 could provide the most efficient way to record data from a multichannel analyzer. In 1965, the 3600 ran a variety of programs to simplify setup and use of the K50, for analyzing data, decomposing spectra into peaks corresponding to certain nuclear excitations, and for driving a digital (Calcomp) plotter. The charge for use of the 3600 computer was around $40K/year, almost $300K/year in 2013 dollars.

The next qualitative change was the introduction of local computing capability in 1967 when the laboratory acquired a Sigma 7 computer, and developed JANUS[213] to take advantage of its flexible time-sharing capability. It was possible, and simple, to set up and record two-dimensional arrays, for example, of particle flight time (or velocity) and energy; one of its first uses was to provide simple mass identification using an Et^2 vs. E plot (E on the x-axis, Et^2 on the y-axis). Et^2 is proportional to mass. Although video displays made it possible to observe and analyze one- and two-dimensional spectra,[214] these spectra were stored on punch cards. Program input was also through punch cards, and running a program meant loading the program, often a deck of a thousand or more cards. Punched cards filled bookshelves in every office.

ADCs, analog to digital converters, convert the voltage of a detector signal, often related to particle energy, to a digital number for use in computer processing and storage. They are a part of almost every experiment in nuclear physics. Access to the outputs of the ADCs we employed was provided by PDP 11-45 mini-computers and CAMAC crates to hold the ADCs. The PDP 11-45s handled input-output and backup to peripheral tape drives, and on-line data analysis. Magnetic disk pack storage provided for additional data and program storage with faster access.

A major step forward occurred in 1980, during the Phase II project when the laboratory purchased a group of VAX computers from the Digital Equipment Corporation: four VAX 750s for data taking and a VAX 780 for more-demanding theoretical computations. Another major step was the gradual conversion to the personal computer scale using the LINUX operating system, begun in 1999 and completed in 2003. Significant, multi-terabyte hard disk storage was provided to permit more efficient analysis of large data sets.

It was only in the late 1970s that one could use terminals for data and program input, and data storage moved to the large nine track tape drives, the consoles one saw in movies of computational facilities until fairly recently. A single experiment might generate over fifty of these tapes.[215] Offices full of punch cards were replaced by closets full of large tapes.

There was then a sequence of different smaller, higher density tape drives: beginning with 8 mm tape in 1990, then DLT digital linear tape, and finally the move in 2000 to taking data on hard disks. Final backup storage is still on tape, but small terabyte disk drives are often used by researchers to take their data back to their home institutions. The closet is now shoebox size.

These increases in computer power and data storage capability had a major effect on the nature of experiments. When computer power was limited and storage slow, only a limited amount of data could be stored. This forced experimenters to set stringent hardware gates (conditions to be satisfied by the data) to limit the input rate to what the computers could handle. It was time consuming, and not always simple, to set sufficiently selective and efficient gates; experiments sometimes failed because a gate had been poorly set. One can now set hardware gates more loosely, record a much richer data set, and set tighter gates after the fact, during off-line data analysis, reducing the chance of failure.

Computer systems and data processing before storage still sometimes limit the rate at which one can take data, especially for complex arrays of detectors, and achieving efficient algorithms for data handling remains a challenge. For example, in the S800, data rates are limited to around 2,000 events/second.

A recent change is the use of digital electronics in many experiments: the time evolution of the signal from detectors is recorded, allowing one to obtain much more detailed information about the detected event, at the cost of much-increased storage requirements. An accompanying advantage is that once one has the basic digital system, much less electronics is required for a particular experiment. It appears that this approach will become dominant in the near future.

CYCLOTRON OPERATIONS

Efficient operation of its accelerators is crucial to the efficiency and reputation of the laboratory. The discussion below is based partly on information provided by Peter Miller and Andreas Stolz. Miller was the first head of the Operations Department and held that position until it was assumed by Stolz in 2008.

During the K50 era (1965 to 1979), accelerator operation during experiments was entirely in the hands of the experimenters. They handled beam tuning for the cyclotron and the beam lines, taking the beam to the experiment, setting up the experimental devices (the Enge spectrometer, for example), changing ion sources when necessary (every few hours when running heavier ions), and, when the machine was running smoothly, napping on a cot set up behind the control console.

They also changed deflector septa in the presence of high radiation levels. This was a motivation to obtain high efficiency beam extraction, since less beam then hit the deflector and it became less radioactive. Even so, one worked at arm's length behind a lead glass shield, and quickly, to minimize radiation exposure. With care, exposure was low, far below prescribed limits.

In 1982, when operation of the K500 was about to begin, a few operators were hired, but faculty often had to take scheduled operation shifts, especially for outside users. By 1986, the operation staff had increased to six plus one supervisor, sufficient to provide single coverage in each shift, of a twenty-four hours per day, seven days per week operating schedule. This was still below the level of two operators per shift that would be necessary to react quickly to the breakdowns common in K500 operation and increase the efficiency of operation. Because of monetary constraints, the situation did not change for some time. For example, in 1995, we were operating the K500 and the K1200, each in standalone mode, with six operators for K1200 experiments, and a single operator for the K500 when it was used to test items for the Coupled Cyclotron Facility.

After K500 operation began, one or two liaison physicists were hired to facilitate interactions of users, especially outside users, with NSCL staff. With the advent of the A1200, we soon learned that efficient tuning of radioactive beams is difficult, requires a high level of insight and experience, and cannot be done by most experimenters. Brad Sherrill, who had built the A1200 and knew it well, devoted a significant fraction of his time to assisting both inside and outside users, often on night shifts. In order to rationalize this situation, a small number of beam physicists was hired to tune and improve the A1200.

A significant upgrade of the Operations Department took place at the beginning of Coupled Cyclotron operation in 2001. Obvious motivations were that there were now two cyclotrons to operate, that at the beginning of operation there would be teething problems, and that the A1900 was a very complex fragment separator. The demands of users were also expected to be more exacting since the easiest radioactive beam experiments had been done and new experiments would employ nuclei with very unusual proton-to-neutron ratios and would require high primary-beam intensity. And it was more expensive to run this larger system, so breakdown time and tuning time had a high value, around $4,000/hour. This provided an incentive to be efficient in all ways. To meet these needs, the group of operators and beam physicists was doubled. Evidence for the wisdom of these moves is that system reliability is still greater than 90 percent, even though the accelerator system is more than twice as complex as previously.

FIGURE 93 (opposite). Control Room for the Coupled Cyclotron Facility (CCF). There are only a few of the knobs, one, in the past (see fig. 34), used to adjust power supplies. And these will soon be replaced by mouse-based controls.

At present (2015), the operator group of twelve is divided into maintenance and operation engineers with different secondary specialties, roughly hardware or software. There are, in addition, five beam physicists who usually have split responsibilities—assistance to users and carrying out their own research programs—three ion source specialists, two vacuum specialists, a five-person technical development group, and a two-person detector lab. The Beam Physicist Group is responsible for the operation and support of experimental devices (S800, beam sweeper, for example) and the A1900 fragment separator. The Technical Development group is responsible for development of new equipment and software for maintaining these devices. The detector lab is charged with developing building, maintaining, and improving detectors used in experiments. At present, Jeff Stetson is the group leader of the Operations Engineers, Jon Bonofiglio: Maintenance Engineers and Vacuum group, Thomas Baumann: Beam Physicists, G. Machicoane: Ion Source group, Andreas Stolz: Technical Development, and John Yurkon: Detector Lab.

During a typical experiment that is in running mode (setup having been completed), two operators and an "Experimenter in Charge" who speaks for the experimental group are always present. A number of individuals from relevant groups that respond to operational problems are "on call": a Beam Physicist, Maintenance Engineer, Ion Source Physicist, Cryogenics Engineer, and Beam Coordinator. These individuals and their contact information are listed on a video screen for easy access; this formality has greatly improved communication in stressful situations.

MAGNETS AND CRYOGENICS
Magnets

The cyclotrons, beam lines, and some of the experimental apparatus at the NSCL depend on magnets that generate or transport the accelerator's beam to experiments and help to characterize the products of experiments. The following discussion is based partly on information provided by Fabio Casagrande, Helmut Laumer, and Al McCartney, in June 2013.

Both room-temperature and superconducting coils are used: room-temperature coils for smaller magnets such as those in the low energy ReA3 beam lines, and superconducting coils for larger magnets, such as those in the A1900 fragment separator. Altogether, the lab, as of June 2013, has 192 magnets with superconducting coils, housed in seventy-four different cryostats. The larger magnets are all superconducting because the cost of electricity for operating a large room-temperature magnet is also large owing to resistance-related losses in the magnet coils.

The wire used in the superconducting coils is a composite of niobium and titanium embedded in a copper matrix and becomes superconducting below 9.2 K. To achieve that temperature, we use a coolant that boils at a lower temperature. The only option is liquid helium, with a boiling point of 4.2 K at atmospheric pressure. Liquid nitrogen with a boiling point of 77 K is used to help cool the coils. The cryogenic system provides the liquid helium and liquid nitrogen needed for magnet operation.

The NSCL superconducting magnets have a complex structure that is different for different magnets. For the ubiquitous beam-focusing quadrupole magnets,[216] the outer room temperature vacuum vessel measures four feet by four feet by two feet. We hang a liquid nitrogen tank and shield from it with low-heat loss links. A helium temperature vessel, containing the magnet iron and superconducting coils, immersed in liquid helium, is suspended from the nitrogen tank by more low-loss links. Both vessels are covered by aluminized tape to reduce heat lost by radiation, and the space between the liquid nitrogen and the room temperature containers is filled with super insulation, many layers of interspersed fiberglass sheets and aluminized Mylar.

This system is remarkably efficient. The beam pipes of the quadrupoles are at room temperature and are four inches in diameter. Only half an inch away the temperature of the helium container is about 4.6 K.[217] The overall heat load on the standard NSCL quadrupole doublets is five liters per day for both helium and nitrogen.

Cryogens

It is well known, in principle, how to produce liquid helium: cool helium gas by (adiabatically) expanding it to reduce the temperature; then forcing the cooled gas at high pressure though a fine nozzle and allowing it to expand into a low-pressure environment. This Joule-Kelvin expansion converts the gas to a liquid. The adiabatic expansion stage is necessary because at higher temperature, a Joule-Kelvin expansion will heat rather than cool helium gas. In practice, the process must be optimized for efficiency and helium liquefiers such as those in the NSCL are complicated and expensive. Liquid nitrogen can be produced by a Joule-Kelvin expansion starting from room temperature, is inexpensive, and the laboratory buys it commercially.

We make our own liquid helium from gas. The gas in a liter of liquid is expensive, around five dollars in 2013, but because it is recycled, the gas cost is relatively small. The cost of the liquid helium is mainly the cost of electricity to run the liquefiers. When the K500 program began, three small liquefiers with a total capacity of thirty-six liters/hour were adequate. We have since had four more generations of liquefiers and since 2011 have been operating two liquefiers with capacities of 800 and 200 liters/hour. These liquefiers, mainly their compressors, consume about 1.4 megawatts at full power. At electricity rates of $100/megawatt-hour, liquid helium costs $0.15/liter.[218] Commercial liquid helium sells for around $7/liter, and is used only for special applications. We also use 33 million liters (1.2 million cubic feet) of liquid nitrogen per year at a cost of $500,000, or $0.015/liter.

Transfer lines lead liquids from the helium liquefier and the liquid nitrogen storage tanks to each group of magnets. For this purpose there are, in 2013, about 620 meters of transfer lines, running everywhere in the laboratory. They are complicated welded and insulated structures, since the helium gas boiled off from the magnets has to be recycled, and some of it is used to cool the gas re-entering the liquefier. Some of the magnets, mostly beam line magnets, are filled in a batch process as needed. Larger magnets—the cyclotrons, the A1900 fragment separator, and the S800 spectrograph—are continuously fed liquid helium.

The liquid nitrogen is not reused, but the boil-off nitrogen has to be carried away from the closed experimental areas, lest it dilute the air and reduce its oxygen content, a danger to humans. For this reason, one sees oxygen monitors everywhere, to provide low-oxygen warnings in case there are accidents that release large quantities of nitrogen or helium.

In principle there is no loss of helium, but in practice, there are always leaks. A much worse loss can result from a long-lasting power failure. The NSCL does not have sufficient helium-gas storage to hold the gas released if all the magnets warm up and the 10,000 liters of liquid helium in their reservoirs evaporate to gaseous helium; such storage would be expensive, and it is more cost-effective at present to accept this rare risk and the subsequent large loss of helium. This choice may change if helium gas becomes more scarce and expensive, shifting the balance of cost toward storage.

New issues will arise in the future. At present we have a small surplus liquefaction capacity, but this will not be adequate if ReA3 is expanded to ReA6 and ReA12 and as extensive testing of FRIB components and then running FRIB begin. A major increase in liquefaction capacity, around a factor of ten, will be required. In addition, systems will be separated, so FRIB/ReA-related testing with its attendant uncertainties does not interfere with reliable operation of the NSCL coupled cyclotrons.

For FRIB there is an additional complication: lower temperatures, around two degrees, are used for the linear accelerator cavities that power the device, in order to increase their efficiency. This is below the so-called lambda point of liquid helium, 2.17 K at atmospheric pressure; the liquid helium has (partly) become a super fluid, and lost all viscosity and resistance to flow. A high level of perfection in welds is required to avoid super leaks.

DESIGN AND MACHINE SHOPS

Owing to its build-it-yourself philosophy, the lab has always had a significant capability in these areas. The common theme is to utilize new capabilities as they become available and to integrate design and machining operations, thereby improving overall efficiency and accuracy in fabrication. This section is partly based upon information provided by Don Lawton and James Wagner, in June 2013.

Mechanical Design

In the early days of the laboratory, nearly all design layouts and documentation were done by paper and pencil on the drafting board. When they became sufficiently powerful, in the early 1980s, we began to use computer-based drawing programs (CAD or computer-assisted design). This led in 1984 to the purchase of a Digital Equipment Corporation VAX computer using an Intergraph-based mechanical design system. This was an expensive system, and the design staff adopted a shift system to better utilize these workstations. The late 1980s/early 1990s saw a change from a central CPU system to desktop UNIX workstations, still utilizing the Intergraph design system, and then to desktop PCs running Bentley MicroStation software. MicroStation was a direct descendant of the Intergraph system that utilized the same user interface and file structure. Also in the 1990s, use of Computer Aided Engineering programs such as COSMOS became more common for solving stress and thermal problems.

In 2004, we purchased the parametric solid modeling program Solid Works and ANSYS engineering products. The ability to envision designs in three dimensions was a quantum change and required a retraining of the staff in solid modeling and parametric design techniques. Then, in 2010, we procured Adept Part Data Management software to manage and protect the thousands of design files that make up our digital drawing inventory.

These computer-based systems take the quality and quantity of design options far beyond what could be done on the drawing board, providing a much higher level of detail and accuracy as the equipment is fabricated. When the construction of the K500 began, a wooden model of the complex structure was built to help designers visualize what they had to design. Now solid modeling

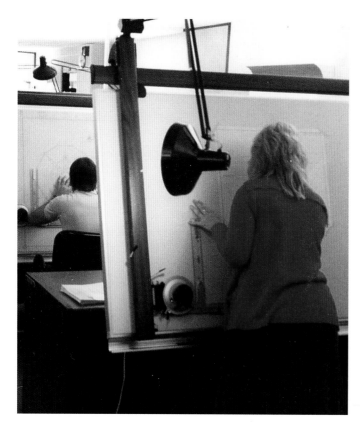

FIGURE 94. Two ways of drafting. *Left*: The pencil and paper approach, up to 1985. *Right*: Don Lawton working with the new CAD system.

programs produce the beautiful and useful renderings that one sees on laboratory walls and desks. The changes are illustrated in figure 94.

Machining

The machine shop has followed a similar trajectory. Prior to 1985, all machining was performed on manual mills and lathes following design documentation on blueprints. The first CNC (Computer Numerical Control) mill, a Hurco BMC 10 machining center with fourth axis capability, was obtained in September of 1985. It employed conversational programming: the machinist interpreted

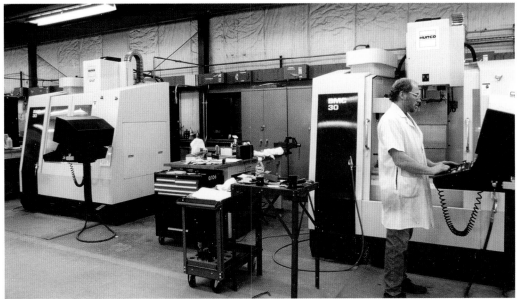

FIGURE 95. Two ways of machining. *Top:* The 1965 version of high tech with fully manual milling machines. *Bottom:* Keith Leslie, recently retired, with the fully automatic machining centers of the present. The input commands are fully digital and one machinist can sometimes operate more than one machine.

the features of the part to be fabricated by inputting necessary (G-code) information to the mill controller. Soon thereafter, another small CNC mill was purchased, and two manual Bridgeport mills were converted to CNC.

In the early 1990s, the machinists were trained in Mastercam CAD/CAM (computer-aided design/ computer-aided manufacturing) software for two, three, and four axis machining; such training has become an ongoing process as software improves. This program generated the G-code information and displayed cutting paths, which greatly improved productivity and the quality of fabricated parts. It allowed the machinist to interface with Mechanical Design drawings and import drawing files, somewhat indirectly, into the CAD/CAM program, eliminating the need for blueprints. Recent purchases of four- and five-axis machines have expanded machining capabilities.

As the CAD/CAM software has evolved, so have the machining centers. Today's machines incorporate more powerful controllers, enhanced milling and turning features, and improved safety and green technology. The machine shop Mastercam CAD/CAM software and Mechanical Design SolidWorks software are compatible with direct file transfer between these programs. These changes are illustrated in figure 95.

Welding is also an important activity, with around six laboratory welders plus some outside contractors (see fig. 96). Cryogenic welding is the major activity of a majority of the welders.

▪ Organization of the Laboratory

The laboratory is at present a complex and interlocking organization. It has three somewhat distinct parts: the FRIB Laboratory, the FRIB Project, and the NSCL Laboratory. The FRIB Laboratory director is Thomas Glasmacher who also serves as the FRIB Project Director. As Project Director he has all authority and responsibility for the construction of FRIB. In the MSU hierarchy, the FRIB Laboratory director is at the organizational level of a dean. The NSCL Laboratory reports to the FRIB Laboratory director.

FRIB LABORATORY

FRIB LABORATORY DIRECTOR: Thomas Glasmacher

DEPUTY LABORATORY DIRECTOR: Paul Mantica

FIGURE 96 (*opposite*). Bob Welton welding parts of the radio frequency system of the K1200 cyclotron. The ability to use a large variety of techniques to weld a variety of metals in different systems is crucial. It is a skill not easily learned but is important to the operation of the laboratory.

ASSOCIATE DIRECTOR FOR USER RELATIONS: Michael Thoennessen

FRIB CHIEF SCIENTIST: Witek Nazarewicz

NSCL LABORATORY

NSCL LABORATORY DIRECTOR: Brad Sherrill

NSCL CHIEF SCIENTIST: Alexandra Gade

ASSOCIATE DIRECTOR FOR OPERATIONS: David Morrissey

ASSOCIATE DIRECTOR FOR EXPERIMENTAL NUCLEAR SCIENCE: Remco Zegers

ASSOCIATE DIRECTOR FOR EDUCATION AND OUTREACH: Artemis Spyrou

THEORETICAL NUCLEAR SCIENCE DEPARTMENT HEAD: Filomena Nunes

ACCELERATOR PHYSICS DEPARTMENT HEAD: Jie Wei

APPLIED ISOTOPE SCIENCE DEPARTMENT HEAD: Georg Bollen

OPERATIONS DEPARTMENT HEAD: Andreas Stolz

MANAGER FOR USER RELATIONS: Jill Berryman

FRIB PROJECT

PROJECT DIRECTOR: Thomas Glasmacher

DEPUTY PROJECT MANAGER: Paul Mantica

ACCELERATOR SYSTEMS DIVISION DIRECTOR: Jie Wei

EXPERIMENTAL SYSTEMS DIVISION DIRECTOR: Georg Bollen

CONVENTIONAL FACILITIES & INFRASTRUCTURE DIVISION DIRECTOR: Brad Bull

JINA-CEE

DIRECTOR JINA-CEE: Hendrik Schatz

■ Appointment Systems

NSCL CONTINUING APPOINTMENT (CA) SYSTEM[219]

In the late 1970s, high-level scientific manpower at the NSCL, mostly faculty members, was becoming inadequate for the tasks at hand: operating a facility whose beams could compete strongly at the international level and performing forefront research with this facility. Tenure slots were not available and it was difficult to hire highly talented individuals in research-only positions because of competition from the higher salaries at National Laboratories and the Research Faculty positions at some other universities, Wisconsin and Washington, for example. MSU did not have such a system.

We proposed that MSU establish a contract tenure system for the NSCL, believing that this would be simpler to achieve than a university-wide system. The proposed system had three ranks: Professor (NSCL), Associate Professor (NSCL), and Assistant Professor (NSCL), with a review and contract-tenure system very similar to the usual academic ranks at MSU. Members of the system could supervise graduate student theses with permission of the student's department. Funding for these positions would come from NSCL grants, and there was a procedure dealing with grant termination or decrease. This system was approved,[220] and first offers were made and accepted in 1981. Offers were limited to individuals who had promise of moving to eventual tenured positions in academic departments at MSU or elsewhere. There were typically four to six such appointments at any time. Two individuals eventually made a transition to the MSU Physics and Astronomy Department. Most of the others eventually moved to senior appointments elsewhere, often at National Laboratories, but played a major role in the laboratory while they were at MSU.

Other NSCL staff members were still appointed in MSU personnel systems that were not always appropriate for the Cyclotron Laboratory or that presented administrative or morale problems. As a result, the NSCL proposed[221] that other senior technical staff be incorporated into the CA system with appropriate titles: Engineer, Staff Engineer and Senior Engineer; Physicist, Staff Physicist, and Senior Physicist; if other specialties arose, e.g., Chemist, similar ranks would be created when more than five individuals were involved. This system was approved by the university on October 10, 1984. At present (2013), there are a total of ninety CA members, roughly equal numbers in the Engineer and Physicist categories and only three in the Professor category. As a result, this system is dominated by a technical rather than an academic culture.

NSCL LABORATORY FACULTY SYSTEM[222]

When MSU was competing for the right to build FRIB, it became clear, as discussed earlier, that we had to increase MSU's faculty-level commitment. Because the NSCL CA system had lost its academic focus, the NSCL proposed a new system, funded with university (not grant) funds made available to the NSCL by the MSU administration, to help attract outstanding faculty. This system had rolling tenure positions. This means that appointments were in a sense term appointments for a fixed period of years, but that after each passing year the term of the appointment remained the same. Of course there were complications related to initial appointment, rank, promotions, etc. Initial appointments were, typically, fixed term for four years and after achieving tenure, for a rolling five years. These individuals held joint appointments in academic departments, mainly in the Physics and Chemistry Departments, at Assistant Professor, Associate Professor, or Professor levels, with the right to supervise graduate students, expectation to teach, and most, but not quite all, typical faculty rights and responsibilities.

There was concern that this system would not be sufficiently attractive, and in a few cases it was not. But the NSCL has, by any measure, attracted truly outstanding individuals to these positions. They have developed exceptionally influential research programs and hold important positions in the NSCL and FRIB: the Associate Director for Experimental Research, the Theoretical Science Department Head, the NSCL Chief Scientist, and the FRIB Accelerator Systems Division Director. At present,[223] nineteen appointments have been made: thirteen in experimental or theoretical physics and six in accelerator physics; one position remains open.

UNIVERSITY DISTINGUISHED PROFESSORS

The University Distinguished Professor rank is selective and is presently held by about 100 individuals, roughly 2 percent of the professorate. They are expected to be leaders in their specialty areas. The NSCL has had a large number of such individuals, reflecting its status as an outstanding research facility at MSU.

1990: Sam M. Austin (Emeritus), Henry Blosser (Emeritus), C. Konrad Gelbke

1997: Walter Benenson (Emeritus), Edwin Kashy (Emeritus)

2003: Brad Sherrill, Gary D. Westfall

2005: David J. Morrissey

2007: Wolfgang Bauer

2009: Thomas Glasmacher

2011: Paul F. Mantica

2013: Georg Bollen

2015: Hendrik Schatz, Michael Thoennessen

JOHN A. HANNAH DISTINGUISHED PROFESSORSHIPS

These professorships are well-funded and are attractive to individuals of the highest reputation and talent. About ten individuals hold these chairs at MSU. The laboratory has benefited greatly from the five Hannah Professors, three of them experimentalists and two theorists, who have spent parts of their careers at the laboratory.

- GEORGE BERTSCH (1971–92, JAHP 1985–92) was a theorist with strong creativity and insight into a great variety of physical phenomena. "His research in nuclear theory began with spectroscopy and particularly giant resonances and went on to the properties of high density matter and their experimental implications. Most recently he has been pursuing the connections between theoretical techniques used in different disciplines . . ."[224] His broad knowledge and willingness to explore new ideas and implications of data with experimenters were exceptional. While at MSU, he developed insights into the nature of heavy-ion collisions and imbedded them in a BUU code that was extremely influential in the study of nucleus-nucleus collisions. He was responsible for the qualitative arguments that led to 200 MeV/nucleon for the proposed energy of the NSCL Coupled Cyclotron Facility.

- DAVID K. SCOTT (1979–87), an experimentalist specializing in the study of heavy-ion–induced collisions, came to MSU from the Lawrence Berkeley Laboratory. He had an unequalled understanding of the science that could be done with beams of heavy ions. Although many at the NSCL were initially doubtful about the value of studying nucleus-induced collisions, his leadership generated an excitement at MSU for heavy-ion research that anticipated future studies of the nuclear medium and of the production of neutron-rich fragments. Scott was an eloquent community spokesman, through his personal interactions, review articles, and talks at major conferences, for the value of heavy-ion research in general and for the MSU/NSCL research in particular.

- P. GREGERS HANSEN (1995–2005) was an experimentalist with a broad range of interests in atomic physics, nuclear structure, and beta decay, among other areas. At the NSCL he was interested in halo nuclei and played a major role in establishing the use of breakup reactions to study the structure of exotic nuclei. Because of his stature and critical mind, he was a force for clear thinking for all of us and especially younger physicists. He died in 2005.

- WOLFGANG MITTIG (2008–present) is an experimentalist with interests in the many physical problems that can be studied with the tools of nuclear science. Most recently his studies have concentrated on the properties of exotic nuclei. He has shown great talent and creativity in the construction of advanced experimental apparatus. While at MSU, he has successfully attacked two challenging problems: a unique high-power production target that can handle the high beam power from FRIB and an Active-Target Time Projection Chamber (A-TPC) that will soon be used for studies of exotic nuclei at the CCF and later at FRIB.

- WITOLD NAZAREWICZ (2014–present) is a nuclear structure theorist and Tom Bonner Prize winner with "broad expertise in nuclear science, interdisciplinary many-body science, and computational physics. His research has made important contributions to nuclear structure and reaction physics, especially in the areas of rare isotopes and nuclear dynamics at the extremes of isospin, mass, and angular momentum."[225] In August 2014, he assumed the position of FRIB Chief Scientist. He will be deeply involved in developing the theoretical infrastructure and expertise that will enhance the impact of the experiments at NSCL and FRIB and our understanding of the nuclear many-body system.

APPENDICES

■ **Appendix A: DOE Review Process for FRIB**

The following are excerpts of the most significant steps in the extensive FRIB process to date as announced by FRIB Project Director Thomas Glasmacher. They are meant to convey the nature of the various review processes.

- SEPTEMBER 1, 2010. The approval of CD-1 establishes the preferred alternative for FRIB and the associated cost and schedule ranges. The projected total project cost for FRIB at CD-1 is $614.5 million with project completion in the first quarter of 2020. We intend to manage the project to an early completion in late 2018.
- MARCH 30, 2011. The U.S. Department of Energy (DOE) Office of Project Assessment and the Office of Nuclear Physics conducted a review of the FRIB Project today. The review committee found that FRIB is making good progress to support the approval of a project baseline (CD-2) in spring 2012 coincident with the approval of civil construction (CD-3A), that management planning is appropriate and that FRIB is addressing prior recommendations. This means the FRIB Project remains on track for scheduled completion.

- OCTOBER 5, 2011. The U.S. Department of Energy Office of Science (DOE-SC) Office of Project Assessment conducted a review of the FRIB Project last week, Sept. 27–29. Charged by the Office of Nuclear Physics, the review committee found that technical design of FRIB is on track to meet its performance expectations, and to establish a performance baseline and commence conventional facilities construction in the spring of 2012. The project was found to be properly managed, including environment, safety, and health aspects, and the review committee recommended to schedule a CD-2/3A review in April 2012.

- APRIL 27, 2012. The U.S. Department of Energy Office of Science (DOE-SC) Office of Project Assessment conducted a review of the FRIB Project this week, April 24–26. Charged by the Office of Nuclear Physics, the review committee assessed all aspects of the FRIB Project—technical, cost, schedule, management, environment, safety, health, and quality assurance. All charge questions to the committee were answered affirmatively at the review closeout session. The successful review confirms FRIB is ready to establish a project baseline for cost, scope, and schedule, and is ready to begin construction of conventional facilities pending approval from the Department of Energy.

- JUNE 6, 2013. The U.S. Department of Energy Office of Science (DOE-SC) assessed FRIB's progress on June 4–5. The two-day review, organized by the DOE-SC Office of Project Assessment, was charged by the DOE-SC Office of Nuclear Physics. Twenty subject matter experts and DOE observers examined the project in five parallel sessions. The committee found that FRIB is making excellent progress, that the project is well-managed, and that it has developed solid and well thought out plans for various funding options and that these plans are realistic and executable. The review committee recommends to proceed with the requested long-lead technical procurement items and to proceed with the CD-2/3A (baseline and start of civil construction) approval process, once the decision for civil construction start is made.

- AUGUST 5, 2013. The DOE Office of Science has released the following statement: "On August 1, 2013, the Department of Energy's Office of Science approved Critical Decision-2 (CD-2), Approve Performance Baseline, and Critical Decision-3A (CD-3A), Approve Start of Civil Construction and Long Lead Procurements, for the Facility for Rare Isotope Beams (FRIB) construction project, which will be located at Michigan State University. As with other DOE Office of Science construction projects, CD-2 formally establishes the cost and schedule for the FRIB project. The Total Project Cost for FRIB is $730M, of which $635.5M will be provided by DOE and $94.5M will be provided by Michigan State University. The project will be completed by 2022. The CD-3A decision allows the project to proceed with long lead procurements. Commencement of the start of civil construction is subject to a Fiscal Year 2014 appropriation."

- JANUARY 23, 2014. On January 22, the Department of Energy Office of Science (DOE-SC) gave the Facility for Rare Isotope Beams (FRIB) official notice that it can now begin construction. This notice completes the process begun when President Obama signed into law the 2014 federal budget approved by the U.S. House of Representatives and the Senate, which includes $55 million to support construction of FRIB in the coming year.

- February 27, 2014. The Michigan Strategic Fund (MSF) board induced [*sic*] more than $90 million for the Facility for Rare Isotope Beams (FRIB) at their regular meeting, held February 25 at FRIB.

- MARCH 17, 2014. Today Michigan State University (MSU) held a groundbreaking ceremony at the site of the future Facility for Rare Isotope Beams (FRIB) on the MSU campus in East Lansing, Michigan. More than 1,000 participants gathered to reflect on MSU's journey to selection by the U.S. Department of Energy Office of Science (DOE-SC) as the site for this new national user facility and also to recommit to the work that lies ahead. Among these participants were members of the Michigan Congressional delegation, representatives from the State of Michigan, leaders from MSU, and representatives from the DOE Office of Science and the DOE-SC Nuclear Physics program.

- JUNE 27, 2014. The DOE Office of Science Office of Project Assessment has concluded its June 24–26 review of the FRIB project and will recommend to the DOE Office of Science Acquisition Executive that FRIB is ready for technical construction (CD-3B). The review committee was organized into eight subcommittees and FRIB staff gave 45 presentations.

- JULY 23, 2014. Today marked a day of solid progress on the conventional facilities front with the first concrete placement for the floor of the linear accelerator tunnel. A team of 24 workers placed the 1,400 cubic yards of concrete, starting around 3 A.M. and finishing around 8 P.M. One-hundred-forty truckloads of concrete were required to complete the task. It was the first of four large concrete placements required to complete the tunnel, which is 1,500 feet long and 70 feet wide.

▪ AUGUST 26, 2014. Today Dr. Patricia Dehmer, the Department of Energy Office of Science acquisition executive for FRIB, has approved *Critical Decision CD-3B: Start of Construction of the Accelerator and Experimental Systems for the Facility for Rare Isotope Beams (FRIB) Project.* With this approval the FRIB Project Team is in an excellent position to finish the FRIB Project in the next six years and to realize the nuclear science community's aspiration for an advanced rare isotope beam facility.

▪ Appendix B. Technical Capability

TECHNICAL EXPERTISE OF NSCL STAFF

▪ RADIATION DETECTION AND SHIELDING
  - Photons
  - Neutrons
  - Protons, heavier ions
  - Interlock systems, safety rated interlock systems

▪ PARTICLE DETECTORS
  - Magnetic spectrographs
  - Scintillation detectors of all kinds
  - Germanium detectors, silicon detectors, diamond detectors
  - Gaseous detectors, including ion chambers, proportional detectors, avalanche detectors, drift chambers
  - Fast-neutron detectors
  - Position-sensitive detectors
  - Time-of-flight detectors

▪ DEEP (SEVERAL MM) IMPLANTATION
  - Radioactive ions
  - Stable ions
  - Absolute counting of atoms and ions

▪ SUPERCONDUCTING MAGNETS AND MATERIALS
  - Accelerator magnets
  - Beam line quadrupoles and dipoles
  - Magnetic spectrometer design

▪ NORMAL TEMPERATURE MAGNETS
  - Design and construction
  - Beam line magnets

▪ PERMANENT MAGNETS
  - Dipoles and multipoles
  - Radiation damage to permanent magnet materials

▪ CRYOGENIC TECHNIQUES
  - Large cryogenic refrigerators, liquid N_2 and He systems
  - Transfer and distribution systems
  - Low helium loss cryostats
  - Cryogenic test dewars for superconducting magnets

- PARTICLE ACCELERATORS, ESPECIALLY CYCLOTRONS AND LINEAR ACCELERATORS
  - Precision sector focused cyclotrons
  - Superconducting cyclotrons
  - Superconducting linear accelerators
  - Orbit dynamics calculations
  - 3-D magnet calculations
  - Beam diagnostic techniques
- BEAM TRANSPORT SYSTEMS
  - Charged particle beam optics
  - Beam diagnostics and simulations from eV to GeV
  - Non-linear beam dynamics, to high order using differential algebra
- ACCELERATORS FOR ONCOLOGY USING SUPERCONDUCTING TECHNOLOGY
  - Neutron producing systems
  - Collimators
  - Proton accelerators
  - Uniform irradiation techniques
  - Beam transport
  - Systems design
- HIGH VACUUM TECHNOLOGY
  - Fore pumps
  - Diffusion pumps, dry scroll, drag, and screw pumps
  - Turbo molecular pumps
  - Leak detection
  - UHV systems, clean room assembly techniques
  - Systems design

- RADIO FREQUENCY SYSTEMS
  - High power (megawatt) systems in the 10 Megahertz range
  - Resonant cavity/circuits design, analysis
  - Signal processing and diagnostics
- POWER SUPPLIES (DC)
  - High power anode supplies for radio frequency systems
  - Multichannel crowbar systems
  - Two and four quadrant precision regulated supplies
  - Power supplies for superconducting magnets
  - Structured controls/interlock systems
  - Filter design
- CONTROL SYSTEMS
  - Distributed real-time supervisory systems
  - Embedded controllers, PLC-based systems
  - Distributed interlock systems
  - GUI user interfaces
  - VME systems, LCSS interfaces
- NETWORKED COMPUTER SYSTEMS
  - FIDDI and Ethernet
  - Distributed and clustered systems
  - Unix systems
- DATA ACQUISITION
  - High speed front ends
  - Networked systems
  - Fiber optics (FIDDI) and Ethernet systems

- ELECTRONIC INSTRUMENTATION
  - Electronic systems architectural design
  - Precision regulation circuits
  - Vacuum, cryogenic instrumentation

- Precision current measurement (pA-kA)
- Application specific instrumentation
- MECHANICAL DESIGN AND CONSTRUCTION (CAD CAM)

TECHNICAL OUTREACH

In addition to the cross-disciplinary research discussed earlier, the NSCL has developed a great amount of technical capability. It has made this expertise available to other research institutions in the U.S. and abroad when it was requested and was possible within the constraints of the NSCL's Cooperative Agreement with the NSF. Some examples:

- ACCELERATOR DEVELOPMENT. Drawings and design information were provided for copies of the K50 cyclotron that were built at Princeton University and at the NASA Lewis Research Center, and for copies of the K500 cyclotron that were built at Texas A&M University (Cyclotron Institute) and Calcutta, India (Variable Energy Cyclotron Center). The main magnet coil for the Texas A&M K500 cyclotron was wound at MSU.

- ACCELERATOR APPLICATIONS. Techniques were developed for production of ^{13}N labeled compounds, and for wear analysis using implanted ^{7}Be ions. Design concepts were developed for accelerators suitable for airport bomb detection, and for use of multipole magnetic fields to achieve uniform charged particle irradiations.[226]

- DATA ACQUISITION. The NSCL data acquisition (NSCLDAQ) and analysis (SpecTcl) systems are used by twenty laboratories. The NSCL data acquisition system was integrated with the GRETINA data acquisition system and used in the first round of GRETINA experiments. A Tcl interface to the EPICS control system, called Epicstcl, is hosted at http://sf.net/projects/epicstcl and has seen over 200 downloads.

- EXPERIMENT PLANNING AND ANALYSIS. LISE^{++} simulates the production of radioactive beams in a variety of reaction types and the effects of a filtering device located downstream of the reaction target. LISE^{++} has become a de facto standard for planning experiments. Available at: http://lise.nscl.MSU.edu/.

- NUSHELLX@MSU, a widely used, state-of-the-art shell model code, is available from brown@nscl.msu.edu. The NSCL website contains a database of programs useful to nuclear physicists.

MSU EXPERIMENTAL APPARATUS DEVELOPMENT

The MSU Cyclotron Laboratory has built or developed an unusually large and varied suite of experimental apparatus, with many at the state-of-the-art of the field at the time. These are listed below.

- Enge Split Pole high resolution spectrograph, using dispersion matching (1.5 keV of 35,000 keV resolution achieved)
- S320 moderate resolution spectrograph
- JANUS, time sharing computer operating system
- Helium jet, for transport of radioactive samples
- Early, possibly first, use of TOF-E for mass identification of light elements
- RPMS, Reaction Products Mass Separator for rare isotope studies
- Superconducting beam lines
- Time-of-flight neutron array for reaction studies
- Beta-NMR end station for studies of magnetic moments
- Beam Swinger for (p, n) reactions, neutron scattering
- Neutron Wall, multi-element neutron detection with gamma discrimination
- Production of ^{13}N and labeled $^{13}N\,^{14}N$ gas for nitrogen fixation and denitrification studies
- Germanium detectors for high-resolution particle detection
- ECR sources with superconducting and normal magnets
- Ninety-two-inch walk-in scattering chamber
- A1200 beam analysis device, fragment separator
- BIG SOL (F. Becchetti, University of Michigan), large superconducting solenoid for reaction product detection
- SUPERBALL (U. Schroeder, Rochester University), large Gd-loaded scintillator for multi-neutron detection
- 4π Array, modular, large-solid-angle, high-rate array
- Miniball, compact, transportable, large-solid-angle array
- BaF2 array for gamma-ray detection
- APEX NaI detectors, modified for directional gamma ray detection
- A1900 fragment separator
- S800, S800 hodoscope

- CRDCs, Cathode Readout Drift Chambers, for beam tracing in S800
- Fifty-three-inch general purpose scattering chamber
- NERO, Neutron Ratio Emission Observer, beta-delayed neutron emission
- Beam Sweeper to separate charged particle from neutrons for LISA/MoNA-LISA
- LISA/MoNA-LISA, multiple detector array for high efficiency detection of high-energy neutrons
- SEETF, Single Event Effect Test Facility for nano-electronics
- LASSA, Large Area Silicon Strip Array for particle spectroscopy
- SEGA, Segmented Germanium Array, for gamma ray spectroscopy
- HIRA, High Resolution Array, for particle spectroscopy
- Beam stopping of fast beams, for stopped beam studies or reacceleration
- RF fragment separator to purify proton rich rare isotope beams
- Plunger detector for nuclear lifetime measurements
- Diamond detectors for high rate particle detection
- LENDA, Low Energy Neutron Detector for reaction studies
- Liquid hydrogen target
- CAESAR, Cesium Iodide Array, for high-efficiency gamma ray detection
- TRIPLEX, Triple Plunger for lifetime measurements
- SUN, Summing NaI spectrometer
- GRETINA (LBNL) gamma ray tracking array
- BECOLA, BEam COoler-LAser spectroscopy for measuring nuclear properties
- Active Target TPC for efficient 3-D detection of ion decays
- Cyclotron beam stopper for reaccelerated beams
- LEBIT, Penning trap for mass measurements
- SIPT, Single-Ion Penning Trap for mass measurements
- Samurai, TPC at RIKEN for equation-of-state studies
- BCS, Beta Counting System for beta decay studies
- (p, p') scattering chamber for cross section studies with rare isotope beams
- Implantation wheel for general implantation studies
- Germanium double-sided strip detectors for beta decay studies
- Bρ-TOF system for mass measurements

- Washington University PET NaI detectors for directional gamma ray detection
- JENSA (ORNL-NSCL) high-density gas jet system

■ Appendix C: A History of the Laboratory Administration

DIRECTORS, CYCLOTRON LABORATORY AND NSCL

Henry G. Blosser (1958–66)

Aaron Galonsky (1966–69)

Henry Blosser (1969–85)

Sam M. Austin and Henry Blosser, Co-directors (1985–89)

Sam M. Austin (1989–92)

C. Konrad Gelbke (1992–2015)

Brad Sherrill (2015–)

DIRECTORS, FRIB LABORATORY

Konrad Gelbke (2008–15)

Thomas Glasmacher (2015–)

CHIEF SCIENTISTS

Brad Sherrill, NSCL and FRIB Chief Scientist (2009–14)

Alexandra Gade, NSCL Chief Scientist (2014–)

Witold Nazarewicz, FRIB Chief Scientist (2014–)

ASSOCIATE DIRECTORS FOR NUCLEAR SCIENCE

B. Hobson Wildenthal (1975–76)

Sam M. Austin (1976–79)

Aaron Galonsky (1979–80)

Walter Benenson (1980–82)

David K. Scott (1982–83)

None (1983–85)

Gary Crawley (1985–87)

Konrad Gelbke (1987–90)

Walter Benenson (1990–95)

David Morrissey (1995–99)

Brad Sherrill (1999–2003)

Michael Thoennessen (2003–07)

Brad Sherrill (2007–09)

David Morrissey (2009–12)

Remco Zegers (2012–)

ASSOCIATE DIRECTORS FOR OPERATIONS

Edwin Kashy (1978–80)

Thomas Glasmacher (2003–09)

Paul Mantica (2009–12)

Daniela Leitner (2012–15)

David Morrissey (2015–)

ASSOCIATE DIRECTOR FOR USERS

Brad Sherrill (2014–15)

Michael Thoennessen (2015–)

RESEARCH DIRECTOR

Sam M. Austin (1983–85)

ASSOCIATE DIRECTOR FOR RESEARCH FACILITIES

Jerry Nolen (1982–92)

ASSOCIATE DIRECTORS FOR ACCELERATOR PHYSICS

Richard York (1993–2010)

Jei Wei (2010–)

ASSOCIATE DIRECTORS FOR EDUCATION

Michael Thoennessen (2007–13)

Remco Zegers (2013–15)

Artemis Spyrou (2015–)

■ Appendix D: External World—MSU

MSU PRESIDENTS

John A. Hannah (1941–69)

Walter Adams (1969–70)

Clifford R. Wharton (1970–78)

Edgar L. Harden (1978–79)

M. Cecil Mackey (1979–85)

John A. DiBiaggio (1985–92)

Gordon E. Guyer (1992–93)

M. Peter McPherson (1993–2004)

Lou Anna K. Simon (2005–)

MSU PROVOSTS

John Cantlon (1969–76)

Clarence L. Winder (1977–86)

David K. Scott (1986–92)

Lou Anna K. Simon (1993–2004)

John Hudzik (*acting*) (2005)

Kim A. Wilcox (2005–13)

June P. Youatt (2013–)

COLLEGE OF NATURAL SCIENCE DEANS)

Richard U. Byerrum (1962–86)

Frank C. Hoppensteadt (1986–96)

George E. Leroi (1996–2007)

R. James Kilpatrick (2007–)

MSU PHYSICS DEPARTMENT CHAIRS

Robert D. Spence (3/16/1950–8/31/1950)

Egon A. Hiedemann (9/1/1950–6/30/1954)

Robert D. Spence (7/1/1954–9/14/1954)

Thomas H. Osgood (9/15/1954–8/31/1955)

Robert D. Spence (9/1/1955–1/11/1956)

Richard Schlegel (1/12/1956–8/31/1956)

Robert D. Spence (9/1/1956–8/31/1957)

Sherwood K. Haynes (9/1/1957–8/31/1969)

Frank J. Blatt (9/1/1969–12/31/1972)

Truman O. Woodruff (1/1/1973–12/31/1975)

William H. Kelly (1/1/1976–9/1/1979)

Julius S. Kovacs (9/1/1979–8/31/1980)

Sam M. Austin (9/10/1980–8/31/1983)

Jack Bass (9/1/1983–8/31/1989)

Gerard M. Crawley (9/1/1989–1/15/1994)

Raymond Brock (8/16/94–2/15/01)

Wolfgang Bauer (2/16/2001–2013)

Phil Duxbury (8/2013–)

■ Appendix E: The External World—NSCL User Committees

NSCL PROGRAM ADVISORY COMMITTEES

NSCL PAC members to date, their institution, and PAC meetings are:

MEMBERS	INSTITUTION	PAC MEETING(S)
Britt, H. C.	LANL	1, 2
Cline, D.	Rochester	1, 2, 3, 4, 5
Koonin, S. E.	Caltech	1, 2, 3, 4, 5, 6
Paul, P.	Stony Brook	1, 2
Scott, D. K.	MSU/NSCL	1, 2, 3
Cramer, J.	University of Washington	3, 4, 5, 6, 7
Viola, V.	Indiana Univesity	3, 4, 5, 6, 7, 8
Benenson, W.	MSU/NSCL	4, 5, 6, 7, 8, 9, 21
Siemens, P.	Texas A&M	5, 6, 7, 8, 9, 10
Stephens, F.	LBNL	8, 9, 10, 11, 12, 13
Vary, J.	Iowa State	6, 7, 8, 9
Young, G.	ORNL	7, 8, 9, 10, 11, 12
Natowitz, J.	Texas A&M	9, 10, 11, 12, 13, 14
Hardy, J.	Chalk River	9, 10, 11, 12, 13, 14
Kashy, E.	MSU/NSCL	10, 11, 12, 13, 14, 15, 16
Randrup, J.	LBNL	10, 11, 12, 13, 14, 15, 16
Haxton, W.	University of Washington	11, 12, 14, 15, 16, 17
Datz, S.	ORNL	13, 14, 15, 16, 18, 19

Henning, W.	Argonne	14, 15, 16, 17, 18, 19, 20, 21
Wozniak, G.	LBNL	14, 15, 16, 17, 18, 19, 20
Mueller, A.	IPN Orsay	15, 16, 17, 18, 19, 20
Crawley, G.	MSU/NSCL	17, 18, 19, 20, 22
Friedman, W.	University of Wisconsin	21, 22, 24
Nazarewicz, W.	ORNL	20, 21, 22, 23, 24, 25, 26
Vandenbosch, R.	University of Washington	21, 22, 23, 24
Janssens, R.	Argonne	22, 23, 24, 25, 26, 27
Austin, S.	MSU/NSCL	23, 24
Geissel, H.	GSI	22, 23, 24, 25, 26, 27
Bollen, G.	MSU/NSCL	25, 26, 27, 28, 29, 30, 31
Das Gupta, S.	McGill	25, 26
Danielewicz, P.	MSU/NSCL	27
Parker, P.	Yale	25, 26, 27, 28, 29
Hjorth-Jensen, M.	Oslo	27, 28, 29, 30, 31, 32
Beene, J.	ORNL	28, 29, 30, 31, 32, 33
Lacey, R.	Stony Brook	28, 29, 30, 31, 32, 33
Kemper, K.	Florida State University	29, 30, 31, 32, 33, 34
Van Duppen, P.	University of Leuven	29, 30, 32, 33, 34, 35
Tribble, R.	Texas A&M	30, 31, 32, 33, 34, 35
Yennello, S.	Texas A&M	31, 32, 33, 34, 35, 36
Mantica P.	MSU/NSCL	32, 33, 35, 36, 37
Dean, D.	ORNL	33, 34, 35, 36, 37, 38
Schatz, H.	MSU/NSCL	34, 38, 39
Aprahamian, A.	University of Notre Dame	34, 35, 36, 37, 38, 39
Wuosmaa, A.	Western Michigan	34, 35, 36, 37, 38, 39
Machiavelli, A.	LBNL	35, 36, 37, 38, 39
Lister, K.	ANL/UMass, Lowell	36, 37, 38, 39
Woods, P.	Edinburgh	36, 37, 38, 39
Natowitz, J.	Texas A&M	37, 38, 39

NSCL EXECUTIVE COMMITTEE MEMBERS

In November 2011, the NSCL Users Group merged with the FRIB User Organization, which has a somewhat different structure. Members served three-year terms, beginning November 1 (earlier two-year terms beginning October 1). Members were elected each year from the general membership of the Users Group; a non-voting liaison representative from MSU was appointed by the Director of the NSCL. Committees selected their own Chair. The NSCL Committees were:

July 1, 1982–Sept. 30, 1982	F. Becchetti	University of Michigan
	A. Galonsky	MSU/NSCL
	J. Huizenga	University of Rochester
	Vic Viola	Indiana University
	G. M. Crawley	MSU/NSCL, *Liaison*
Oct. 1, 1982–Sept. 30, 1983	F. Becchetti	University Michigan, *Chair*
	J. Kolata	University of Notre Dame
	V. Viola	Indiana University
	D. Youngblood	Texas A&M University
	A. Galonsky	MSU/NSCL, *Liaison*
Oct. 1, 1983–Sept. 30, 1984	J. Kolata	University of Notre Dame, *Chair*
	F. Prosser	University of Kansas
	R. Tickle	University of Michigan
	D. Youngblood	Texas A&M University
	A. Galonsky	MSU/NSCL, *Liaison*
Oct. 1, 1984–Oct. 31, 1985	J. Kolata	University of Notre Dame
	L. Lee	SUNY, Stony Brook
	F. Prosser	University of Kansas
	R. Tickle	University of Michigan, *Chair*
	A. Galonsky	MSU/NSCL, *Liaison*

Nov. 1, 1985–Oct. 31, 1986	D. Kovar	ANL
	L. Lee	SUNY, Stony Brook
	F. Prosser	University of Kansas, *Chair*
	R. Tickle	University of Michigan
	A. Galonsky	MSU/NSCL, *Liaison*
Nov. 1, 1986–Oct. 31, 1987	D. Kovar	ANL, *Chair*
	K. Kwiatkowski	Indiana University
	L. Lee	SUNY, Stony Brook
	J. Saladin	University of Pittsburgh
	A. Galonsky	MSU/NSCL, *Liaison*
Nov. 1, 1987–Oct. 31, 1988	D. Kovar	ANL, Chair
	K. Kwiatkowski	Indiana University
	J. Saladin	University of Pittsburgh
	L. Sobotka	Washington University–St. Louis
	A. Galonsky	MSU/NSCL, *Liaison*
Nov. 1, 1988–Oct. 31, 1989	T. Awes	ORNL
	K. Kwiatkowski	Indiana University
	J. Saladin	University of Pittsburgh
	L. Sobotka	Washington University, *Chair*
	A. Galonsky	MSU/NSCL, *Liaison*
Nov. 1, 1989–Oct. 31, 1990	T. Awes	ORNL
	A. Nadasen	University of Michigan
	L. Sobotka	Washington University, *Chair*
	G. Wozniak	LBNL
	A. Galonsky	MSU/NSCL, *Liaison*

Nov. 1, 1990–Oct. 31, 1991	T. Awes	ORNL
	J. Kolata	University of Notre Dame
	A. Nadasen	University of Michigan
	G. Wozniak	LBNL, *Chair*
	A. Galonsky	MSU/NSCL, *Liaison*
Nov. 1, 1991–Oct. 31, 1992	J. Kolata	University of Notre Dame
	A. Nadasen	University of Michigan
	U. Schroeder	University of Rochester
	G. Wozniak	LBNL, *Chair*
	A. Galonsky	MSU/NSCL, *Liaison*
Nov. 1, 1992–Oct. 31, 1993	F. Bertrand	ORNL
	J. Kolata	University of Notre Dame, *Chair*
	U. Schroeder	University of Rochester
	R. Vandenbosch	University of Washington
	N. Anantaraman	MSU/NSCL, *Liaison*
Nov. 1, 1993–Oct. 31, 1994	F. Bertrand	ORNL
	U. Schroeder	University Rochester, *Chair*
	R. Vandenbosch	University of Washington
	M. Wiescher	University of Notre Dame
	N. Anantaraman	MSU/NSCL, *Liaison*
Nov. 1, 1994–Oct. 31, 1995	F. Bertrand	ORNL
	R. Boyd	Ohio State University, *Chair*
	R. Vandenbosch	University of Washington
	M. Wiescher	University of Notre Dame
	N. Anantaraman	MSU/NSCL, Liaison

Nov. 1, 1995–Oct. 31, 1996	R. Boyd	Ohio State University, *Chair*
	K. Kemper	Florida State University
	R. Warner	NSCL/Oberlin College
	M. Wiescher	University of Notre Dame
	N. Anantaraman	MSU/NSCL, *Liaison*
Nov. 1, 1996–Oct. 31, 1997	R. Boyd	Ohio State University
	K. Kemper	Florida State University, *Chair*
	W. Loveland	Oregon State University
	R. Warner	NSCL/Oberlin College
	N. Anantaraman	MSU/NSCL, *Liaison*
Nov. 1, 1997–Oct. 31, 1998	R. Charity	Washington University
	K. Kemper	Florida State University, *Chair*
	W. Loveland	Oregon State University
	R. Warner	NSCL/Oberlin College
	N. Anantaraman	MSU/NSCL, *Liaison*
Nov. 1, 1998–Oct. 31, 1999	R. Charity	Washington University
	W. Loveland	Oregon State University, *Chair*
	G. Peaslee	Hope College
	A. Wuosmaa	ANL
	N. Anantaraman	MSU/NSCL, *Liaison*
Nov. 1, 1999–Oct. 31, 2000	R. Charity	Washington University, *Chair*
	G. Peaslee	Hope College
	R. Varner	ORNL
	A. Wuosmaa	ANL
	N. Anantaraman	MSU/NSCL, *Liaison*

Nov. 1, 2000–Oct. 31, 2001	G. Peaslee	Hope College, *Chair*
	R. Varner	ORNL
	J. Winger	Mississippi State University
	A. Wuosmaa	ANL
	N. Anantaraman	MSU/NSCL, *Liaison*
Nov. 1, 2001–Oct. 31, 2002	A. Aprahamian	University of Notre Dame
	P. Cottle	Florida State University
	R. Varner	ORNL
	J. Winger	Mississippi State University
	N. Anantaraman	MSU/NSCL, *Liaison*
Nov. 1, 2002–Oct. 31, 2003	A. Aprahamian	University of Notre Dame
	P. Cottle	Florida State University
	D. Shapira	ORNL, *Chair*
	J. Winger	Mississippi State University
	N. Anantaraman	MSU/NSCL, *Liaison*
Nov. 1, 2003–Oct. 31, 2004	A. Aprahamian	University of Notre Dame
	P. Cottle	Florida State University
	D. Shapira	ORNL, *Chair*
	W. Walters	University of Maryland
	N. Anantaraman	MSU/NSCL, *Liaison*
Nov. 1, 2004–Oct. 31, 2005	P. DeYoung	Hope College
	D. Shapira	ORNL, *Chair*
	W. Walters	University Maryland
	I. Wiedenhoever	Florida State University
	N. Anantaraman	MSU/NSCL, *Liaison*

Nov. 1, 2005–Oct. 31, 2006	P. DeYoung	Hope College
	W. Walters	University of Maryland
	J. J. Ressler	Simon Fraser University
	I. Wiedenhoever	Florida State University
	N. Anantaraman	MSU/NSCL, *Liaison*

Nov. 1, 2006–Oct. 31, 2007	P. DeYoung	Hope College
	I.-Y. Lee	LBNL
	R. Grzywacz	University of Tennessee
	I. Wiedenhoever	Florida State University
	N. Anantaraman	MSU/NSCL, *Liaison*

Nov. 1, 2007–Oct. 31, 2008	J. Blackmon	Louisiana State University
	R. Grzywacz	University of Tennessee
	I.-Y. Lee	LBNL, *Chair*
	S. Tabor	Florida State University, *Chair*
	N. Anantaraman	MSU/NSCL, *Liaison*

Nov. 1, 2008–Oct. 31, 2009	J. Blackmon	Louisiana State University
	J. Cizewski	Rutgers University
	I.-Y. Lee	LBNL, *Chair*
	S. Tabor	Florida State University
	N. Anantaraman	MSU/NSCL, *Liaison*

Nov. 1, 2009–Oct. 31, 2010	J. Blackmon	Louisiana State University, *Chair*
	J. Cizewski	Rutgers University
	M. Famiano	Western Michigan University
	S. Tabor	Florida State University
	N. Anantaraman	MSU/NSCL, *Liaison*

Nov. 1, 2010–Oct. 31, 2011	J. Cizewski	Rutgers University, *Chair*
	P. Fallon	LBNL
	M. Famiano	Western Michigan University
	K. Rykaczewski	ORNL
	N. Anantaraman	MSU/NSCL, *Liaison*

■ Appendix F: The External World—National Organizations

NSAC CHAIRS

William A. Fowler	1977–79	Caltech
Herman Feshbach	1979–82	MIT
John P. Schiffer	1983–85	ANL
Ernest Henley	1986–89	University of Washington
Peter Paul	1989–91	SUNY, Stony Brook
Ernie Moniz	1992–95	MIT
R. G. Hamish Robertson	1995–96	University of Washington
C. Konrad Gelbke	1997–99	MSU/NSCL
James Symons	2000–02	LBNL
Richard. F. Casten	2003–05	Yale University
Robert Tribble	2006–08	Texas A&M
Susan J. Seestrom	2009–11	LANL
Donald F. Geesaman	2012–14	ANL

APS-DIVISION OF NUCLEAR PHYSICS CHAIRS

MSU/NSCL members are listed in **boldface**.

L. I. Schiff	G. C. Phillips	J. P. Schiffer
H. H. Barschall	T. Lauritsen	S. S. Hanna
M. Goldhaber	F. Ajzenberg-Selove	J. Weneser
H. Feshbach	B. L. Cohen	P. Axel

E. M. Henley

G. T. Garvey

S. M. Austin

G. F. Bertsch

E. Hayward

L. Rosen

R. G. Stokstad

E. G. Adelberger

S. E. Koonin

R.A. Eisenstein

J. Ball

G. M. Crawley

W. C. Haxton

N. Benczer-Koller

C. B. Dover

J. D. Walecka

L. L. Riedinger

B. C. Clark

S. Freedman

W. Henning

R. G. H. Robertson

J. Moss

C. Glashausser

A. B. Balantekin

D. F. Geesamen

B. M. Sherrill

S. J. Seestrom

R. G. Millner

R. F. Casten

L. S. Cardman

W. A. Zajc

R. E. Tribble

R. D. Mckeown

B. Mueller

A. Aprahamian

John F. Wilkerson

Gordon Cates

M. Thoennessen

NOTES

1. In October 1977, the NSF and ERDA (now DOE) formed the Nuclear Science Advisory Committee (NUSAC and later, NSAC) to advise the agencies on priorities in nuclear science. It was chaired by W. A. Fowler of Caltech during 1978 and 1979. NSAC played an important role in the future evolution of the NSCL into the FRIB era.

2. Public exposure to radiation from FRIB has been carefully limited and evaluated. It easily satisfies MSU's conservative ALARA standards of ten mrem annually, and is much smaller than that from environmental and medical backgrounds. See: R. M. Ronningen, personal communication. The amount is extremely small because the facility is deep underground and because massive shielding of iron and concrete surrounds the points that produce intense radiation. For comparison, according to EPA estimates, the average dose received in the U.S. is 620 mrem each year. The additional radiation dose one receives living in Denver rather than Lansing for a year is about fifty-five mrem. Examples of other exposures are: lumbar spine X-ray: 150 mrem; lumbar spine CT scan: 600 mrem; mammogram: 40 mrem. See: F. A. Mettler et al., *Radiology* 248 (2008): 254.

3. MSU had a variety of names as it grew from an institution involved mainly in agricultural education to a major research university. Some of the details of MSU's evolution, and the accompanying evolution of its name, are described in the next section.

4. John A. Hannah, *A Memoir* (East Lansing: Michigan State University Press, 1980): 37.

5. E. Rutherford, *The Philosophical Magazine*, series 6, vol. 21. (May 1911): 669.

6. In a tandem Van de Graaff, or more simply, a Tandem, that voltage is used twice. The terminal has a positive voltage, and a negatively charged ion is accelerated by the attraction of the positively charged terminal. In the terminal, the ion passes through a thin foil, its excess electrons are stripped away, it becomes positive, and is now repelled by the positive terminal. Many tandem accelerators were built by commercial companies and purchased for use in nuclear physics laboratories, with the first U.S. tandem installed at the University of Wisconsin in 1959–60.

7. See: www.phy.ornl.gov/hribf/accelerator/tandemweb/.

8. I did my thesis experiments in 1959 on a 4 MeV Van de Graaff at the University of Wisconsin.

9. E. O. Lawrence and M. S. Livingston, *Phys. Rev.* 40 (1932): 19. The invention was patented on Feb. 20, 1934.

10. H. A. Bethe and M. E. Rose, "The Maximum Energy Obtainable from the Cyclotron," *Phys. Rev.* 52 (1937): 1254; 22 MeV protons were produced in the ORNL eighty-six-inch cyclotron: J. A. Martin, "Radioisotope Production in the ORNL 86-Inch Cyclotron," *Phys. Rev.* 91 (1953): 224; B. L. Cohen and R. V. Neidigh *Phys. Rev.* 93 (1954): 282.

11. In its later years, it was mainly used for pioneering treatments in cancer therapy.

12. Proceedings of the Sixth International Cyclotron Conference, Vancouver, Canada, 18–21 July 1972 (New York: Am. Inst. Physics, 1972). Cyclotron information compiled by F. T. Howard.

13. Proceedings of the Eighth International Conference on Cyclotrons and Their Applications, Bloomington, IN, 18–21 September 1978. Cyclotron information compiled by John A. Martin.

14. Madison Kuhn, Official Historian of Michigan State University, *Michigan State: The First One Hundred Years* (East Lansing: Michigan State University Press, 1955).

15. Clarence Suelter, "The College of Natural Science at Michigan State University 1855–2005" (East Lansing: College of Natural Science, Michigan State University, 2005), as a CD-ROM.

16. Madison Kuhn, op. cit, 1.

17. Madison Kuhn, op. cit., 9.

18. Madison Kuhn, op. cit., 9 and 10.

19. Its official title was: "An Act Donating Public Lands to the Several States and Territories which may provide Colleges for the Benefit of Agriculture and the Mechanic Arts."

20. Clarence Suelter, op. cit., 24

21. John A. Hannah, *A Memoir*, op. cit, 95.

22. Shanghai rankings: www.shanghairanking.com/SubjectEcoBus2012.html; for other rankings, see: www.msu.edu/about/rankings-and-recognitions/index.html.

23. The planning committee consisted of Egon E. Hiedemann, C. D. Hause, and R. D. Spence.

24. Report of Physics and Astronomy Department activities, dated 1959. In 1959, research in the Physics Department involved experimental work in nuclear magnetic resonance, infrared spectroscopy, ultrasonics, solid-state physics, low-temperature physics, properties of metals, accelerator design, high-energy nuclear physics, and low-energy nuclear physics. There was theoretical research on relativity, philosophy and the history of physics, the theory of waves, statistical mechanics, elementary particles, solid state physics, non-linear field theory, the theory of spectra, and astronomy.

25. M. Muelder in letter to Richard Schlegel, dated February 10, 1978.

26. Richard Schlegel in a letter to Henry Blosser, dated December 14, 1977.

27. Lawrence Jones et al., *Innovation Was Not Enough: A History of the Midwestern Universities Research Association (MURA)* (Hackensack, NJ: World Scientific Publishing Co., 2010).

28. Undated (probably 1955) minutes of MSU Department of Physics Meeting: Dean Osgood with the approval of Dean Muelder applied for membership in MURA.

29. The website available at www.psl.wisc.edu/aboutpsl/history provides a brief history.

30. See an account of Ballam's career: http://news.stanford.edu/news/1999/june16/memballam-616.html.

31. M. Muelder in a letter to Richard Schlegel, dated February 10, 1978.

32. This commentary is based on the minutes of these two meetings prepared by Physics Department Faculty Secretary J. Stephen Kovacs.

33. "A Proposed Heavy-Ion Cyclotron," Report of the Committee on Nuclear Research, June 1957.

34. From an interview with J. S. Kovacs, conducted by S. Austin, July 27, 2012.

35. Transcription of Interview of Julius Kovacs on September 19, 2001. Oral History Project, Michigan State University Archives and Historical Collections, East Lansing, MI.

36. Transcription of Interview of Henry Blosser on August 9, 2000. Oral History Project, Michigan State University Archives and Historical Archives, East Lansing, MI.

37. Letters from B. L. Cohen and R. S. Livingston, dated January 14 and 17, 1958, resp.

38. Haynes's offer letter to Blosser, dated January 21, 1958.

39. Transcript of Interview of Henry Blosser, op. cit.

40. A discussion of this possibility began, after preliminary conversations, in a letter from Blosser to Bromley, dated May 14, 1958. Bromley expressed interest in a four-page letter to Haynes, dated May 27,

1958. A letter from Blosser to Bromley, dated September 2, 1958, concerns an offer to Bromley (no record available).

41. Letter from Bromley to Haynes, dated October 1, 1958.

42. Handwritten letter from Blosser to Bromley, dated October 5, 1958. Blosser states that this offer has President Hannah's approval but that it seems better to keep it informal at this point.

43. Research at MSU has been supported by several federal agencies over the years. The most important of these were the National Science Foundation (NSF) and subsets of three larger agencies with somewhat different, and evolving, overall functions: the Atomic Energy Commission until 1975, the Energy Research and Development Administration (ERDA) from 1975 to 1977, and the Department of Energy (DOE) from 1977 on.

44. H. A. Bethe and M. E. Rose, "The Maximum Energy Obtainable from the Cyclotron," *Phys. Rev.* 52 (1937): 1254.

45. H. G. Blosser et al., *Rev. Sci. Instr.* 29 (1958): 819.

46. Morton M. Gordon accepted the MSU position on March 18, 1959, to arrive by July 1. Gordon was slowly becoming blind. By 1965, he could only see inch-high characters in black crayon, and he soon became functionally blind. He was, nevertheless, still able to carry out complex calculations, and was instrumental in developing techniques for the design of the MSU cyclotrons and the Indiana University cyclotron. He was helped by his wife Bernice who read technical material, including equations, to him. He published prolifically and taught until the end of his career. He was a popular lecturer, and wrote in chalk on blackboards whose panels he delineated by weighted strings.

47. Letter from Paul W. McDaniel, Acting Director of the Division of Research at DOE, to MSU Vice President Milton Muelder, dated October 29, 1959.

48. Letter from V. P. Muelder to President Hannah, dated November 2, 1961.

49. Letter from P. W. McDaniel, Acting Director, AEC Division of Research, to M. Muelder, dated October 29, 1959.

50. See article entitled "MISTIC" on Wikipedia.org and cites therein.

51. Details on these events are outlined in a letter, from Henry Blosser to Vice President Phil May of MSU, dated June 3, 1960.

52. Specifically, for 40 MeV protons, a current of 2.8 ma is predicted with energy spread of 30 keV, and spot size divergence (full width, full angle) of 4 milli-radian-cm radially and 20 milli-radian-cm axially. Each of these figures is an order of magnitude, or more, better than typical performance of present cyclotrons.

53. Letter to President Hannah from Milton E. Muelder, Vice President and Dean, MSU, dated March 20, 1961.

54. Letter to J. Howard McMillan, NSF Program Director for Nuclear Physics, from Milton E. Muelder, Vice President and Dean, MSU, dated April 19, 1961.

55. Letter from George Kolstad (DOE) to MSU Vice President Milton Muelder, dated September 29, 1961.

56. Award Letter from Alan T. Waterman, Director NSF, to President Hannah, dated October 4, 1961.

57. Progress Report "Cyclotron Laboratory Facts and Photographs," prepared by Henry Blosser, July 1963.

58. Chief Engineer Al Schulke; Designers: Guenther Stork, Dick Dickenson; Technicians: Bill Harder, Hans Rothmann; Computer Programmers: David Johnson, Wescott; Machinists: Norval Mercer, Jack Kittsmiller, Floyd Wagner. Stork and Johnson played important roles in cyclotron design construction through all generations of laboratory cyclotrons.

59. Proposal to the NSF for Operating Support and Additional Experimental Equipment for the MSU Cyclotron, July 1963 (Grant of $235,000 for calendar 1964 operation and $389,000 for ancillary equipment).

60. Proposal to the NSF for "Support of the Nuclear Physics Program for the MSU Cyclotron," June 1964; $450,000 requested; MSU support of $332,000. Grant of $430,000 for calendar 1965 operation.

61. Proposal to NSF for "Support of the Nuclear Physics Program for the MSU Cyclotron," July 1965, 13.

62. Letter from Henry Blosser to President Hannah, dated November 21, 1963.

63. Proposals to the NSF for: "Support of the Nuclear Physics Program for the MSU Cyclotron," July 1965 (Requested $3,380,000, for five years beginning at $550,000 in 1966, increasing to $760,000 in 1970. Grant of $567.400 for calendar year 1966). "Data Processing System for MSU Cyclotron Laboratory" (Requested $355,000, Grant of $355,000 for Mar. 21, 1966 to Mar. 21, 1967).

64. L. Kull and E. Kashy, *Phys. Rev.* 167 (1968): 963.

65. Kull has had a remarkable career, becoming a founder, then President and COO of Science Applications International, a major consulting firm.

66. Following a design of G. Danby, BNL.

67. P. J. Plauger became a well-known author and entrepreneur; he founded the software companies Whitesmiths and Dinkumware.

68. J. O. Kopf and P. J. Plauger, "JANUS: A Flexible Approach to Real-Time Timesharing," Proceeding AFIPS, 1968, Fall Joint Computer Conference, part II: 1033–42.

69. W. Benenson, R. A. deForest, R. F. Au, D. L. Bayer, and W. E. Merritt, IEEE Transactions N-16 (1969): 145.

70. H. G. Blosser et al., "Ultra-High Resolution System for Charged Particle Studies of Nuclei," *Nucl. Instr. Meth.* 91 (1971): 61.

71. J. A. Nolen Jr. and R. J. Gleitsmann, "High-resolution study of Ca40 via inelastic proton scattering at 35 MeV," *Phys. Rev. C* 11 (1975): 1159.

72. S. M. Austin and G. M. Crawley, eds., *Proc. Symp. Two Body Force in Nuclei* (New York: Plenum Press, 1972).

73. P. J. Locard, S. M. Austin, and W. Benenson, "The Effective Interaction and the ^{7}Li(p, p') ^{7}Li(478 keV) and ^{7}Li(p, n) ^{7}Be(431 keV) Reactions," *Phys. Rev. Lett.* 19 (1967): 1141.

74. Henry Blosser, quoted in Minutes of the MSU Physics Dept. Advisory Committee, January 27, 1971.

75. Letter from R. T. Siegel to Henry Blosser, dated August 7, 1967.

76. Letters from Robert Wilson to Henry Blosser, dated July 24, 1967 and January 1, 1968, and from Henry Blosser to Wilson dated January 17, 1968.

77. G. M. Crawley, E. Kashy, W. Lanford, and H. G. Blosser, "High Resolution Study of the Particle-Hole Multiplets in ^{208}Bi," *Phys. Rev.* 8 (1973): 2477.

78. 1968–69 Annual Report of the MSU Cyclotron Laboratory.

79. "Proposal for a Versatile Trans-Uranic Research Facility Utilizing a 720(Z^2/A) MeV Variable Energy Multi-particle Cyclotron" dated May 1969, plus July 1969 addenda.

80. Letter to Glenn T. Seaborg from Provost John E. Cantlon and President Walter Adams, dated August 1, 1969.

81. Letter from Paul W. McDaniel, Director of Division of Research, AEC, to John E. Cantlon, dated April 2, 1970.

82. Letters from Henry Blosser to George Regosa, dated June 24, 1969, and August 5, 1969.

83. S. E. Woosley, A. Heger, T. Rauscher, and R. D. Hoffman, *Nucl. Phys. A* 718 (2003): 3c.

84. Cary N. Davids, Helmut Laumer, and Sam M. Austin, "Production of ^{11}B and ^{10}B by Proton Spallation of C^{12}," *Phys. Rev. Lett.* 22 (1969): 1388.

85. S. M. Austin, A. Galonsky, J. Bortins, and C. P. Wolk, "A Batch Process for the Production of ^{13}N ^{14}N Gas," *Nucl. Instr. Meth.* 126 (1975): 373.

86. C. P. Wolk et al., "Pathway of nitrogen-metabolism after fixation of N-13-labeled nitrogen gas by cyanobacterium, anabaena-cylindrica," *J. Biol. Chem.* 251 (1976): 5027.

87. See a compilation of relevant papers in a letter from James M. Tiedje to Henry Blosser, dated September 17, 1981.

88. Interview of Robert Doering by Orville Butler on December 9, 2008, Niels Bohr Library & Archives, American Institute of Physics, College Park, MD.

89. Princeton University (completed in 1969) and NASA Lewis Research Center (first beam July 1972).

90. See Annual Reports of the MSU Cyclotron Laboratory from this period.

91. Physics Survey Committee, Physics in Perspective, Washington, DC, National Academy of Sciences, 1972. Report of the Ad Hoc Panel on Heavy Ion Facilities, Washington, DC, National Academy of Sciences, 1974. This committee, chaired by Herman Feshbach of MIT, met in July 1973.

92. Proceedings of the Sixth International Cyclotron Conference, Vancouver, Canada, July 18–21, 1972 (New York: Am. Inst. Physics, 1972): 24.

93. H. G. Blosser at the Dedication of the NSCL, Wharton Center, MSU, September 27, 1982.

94. Superconducting Heavy Ion Cyclotron, C. B. Bigham, J. S. Fraser, and H. R. Schneider, AECL4654.

95. Historical Outline of MSU Superconducting Cyclotron Program, H. G. Blosser.

96. "Proposal for a Prototype Superconducting Magnet for a Heavy Ion Cyclotron," MSUCP-28, July 1974.

97. The entire system consisted of a superconducting coil wound with niobium-titanium superconducting wire; a cryostat or vacuum bottle to hold the liquid helium that cooled the wire and insulate the cold coil from the room temperature surround; the iron necessary to shape the magnetic field and confine it to the region of the cyclotron; and a refrigeration system to provide the liquid helium to cool the coil.

98. This was a large dipole construction project for the joint U.S.-USSR magneto-hydrodynamic program.

99. H. G. Blosser and D. A. Johnson, "Focusing Properties of Superconducting Cyclotron Magnets," Nucl. Instr. Meth. 121 (1974): 301.

100. Merritt Mallory received his PhD in Physics from MSU in 1966. He then went to ORNL where he worked on the development of cold cathode ion sources for cyclotrons. He returned to MSU in a Specialist position and was involved in Penning ion source development for the K50 and K500 cyclotrons and for installation and testing of the K500 cyclotron.

101. Letter dated March 12, 1969 from W. C. Parkinson on AUA Letterhead reporting a January 11, 1969, meeting at O'Hare.

102. Letter from Henry Blosser to University Nuclear Scientist Neighbors, dated October 4, 1974.

103. Letter from Henry Blosser to MSU Provost John Cantlon, dated October 23, 1974.

104. Letter from Kirk W. McVoy of the University of Wisconsin to fifteen Midwestern nuclear physicists, dated November 25, 1974.

105. Letter from Paul Quin of the University of Wisconsin to Henry Blosser, dated December 17, 1974.

106. Letter from Kirk McVoy and Paul Quin to Members of the Midwest Nuclear Physics Community, dated February 21, 1975.

107. Letter from Henry Blosser to the Midwest Nuclear Group, dated July 11, 1975.

108. Meeting report, letter from Henry Blosser for Midwest Nuclear Group, dated October 27, 1975.

109. From undated draft showing the signatures received.

110. MSUCL-222 "Proposal to NSF for a National Facility for Research with Heavy Ions using Coupled Superconducting Cyclotrons," dated September 1976.

111. Letter from MSU President Clifton R. Wharton Jr. to NSF, dated August 30, 1976.

112. The original title was NUSAC, but that was found to be in use by another organization. In this document, we will always use NSAC for this organization.

113. Other members of the Milan team were Elsa Fabrici and Giovanni Bellomo, who were at the Cyclotron Laboratory through 1980. From early 1980 until his death in February 1984, Resmini was often on leave to further the Milan K800 project.

114. MSUCL-222-A* "Proposal for a National Facility for Research with Heavy Ions using Coupled Superconducting Cyclotrons," Feb. 1978, prepared for the NSAC Subcommittee. It was an update of the 1976 proposal to NSF.

115. "The Future of Nuclear Science," National Academy of Sciences (1977), G. Friedlander, Chair; "Physics in Perspective, Vol. II," National Academy of Sciences (1972).

116. Henry Blosser in MSU Project History, dated April 1982.

117. MSUCL-282, Conceptual Design Report for Phase II of a National Superconducting Cyclotron Laboratory for Research with Heavy Ions, dated December 1978.

118. Telephone call from Senator Levin's office, 4:05 P.M., January 22, 1980.

119. Letter to Henry Blosser from DOE Chicago Operations Office, dated September 29, 1982.

120. Dated November 7, 1979.

121. Oral communication from DOE representatives to Robert Tribble.

122. From H. G. Blosser's Vitae, March 2000. Initial NSF grant was for $400,000, July 1977 to July 1980; followed by grants of $700,000, March 1978 to December 1980, and $324,460, June 1979 to December 1980.

123. Memoranda to lab staff and to NSCL Users Group, from Henry Blosser, Oct. and Nov. 1981.

124. M. L. Mallory, "Initial operation of the MSU/NSCL superconducting cyclotron," *IEEE Trans. Nucl. Sci.* 30 (1983): 2061.

125. MSU Board of Trustees minutes, June 23 and 24, 1977.

126. J. Berryman, private communication.

127. Memo from David K. Scott, Provost to Austin and Blosser, December 7, 1987.

128. Letter to lab staff from R. Dickenson.

129. During a university-wide budget crunch in 1981, the Department of Astronomy and Astrophysics was threatened with dissolution. The Department of Physics, after significant controversy, proposed to "assume the administration of the present faculty and resources of the Department of Astronomy and Astrophysics." This was accepted by the MSU administration and, in 1981, Physics became Physics and Astronomy.

130. NSCL Users Newsletters of March 1983, January 1984, and August 1984.

131. G. D. Westfall et al., "Light particle spectra from 35 MeV/nucleon ^{12}C-induced reactions on ^{197}Au," *Phys. Rev. C* 29 (1984): 861.

132. The S320 is described in NSCL Annual Report for 1979–80: J Nolen et al., "Design and Construction of a K = 320 Spectrograph for Phases I," 72–73. See also: B. M. Sherrill, PhD thesis, Michigan State University, 1984, which lists other contributors: W. Benenson, S. Bricker, W. Lynch, J. van der Plicht, D. Swan, J. Winfield, and J. Yurkon.

133. L. H. Harwood and J. A. Nolen, "A Reaction-Product Mass Separator for Energetic Particles at MSU/NSCL," *Nucl. Instr. Meth.* 186 (1981): 435. See also NSCL Annual Reports from this era.

134. A qualitative description of Penning and ECR sources is given by R. Scrivens, "Electron and Ion Sources for Particle Accelerators," available at: https://cds.cern.ch/record/941321/files/p495.pdf.

135. Peter Miller, private communication. The best lifetimes were obtained with Hf cathodes and nitrogen support gas.

136. P. Apard et al., "Production of Multiply Charged Xenon Ions" *Physics Letters* 44A (1973): 432.

137. T. A. Antaya and Z. Q. Xie, in Proceedings of the Seventh Workshop on ECR Ion Sources (Juelich, W. Germany, 1986), 72.

138. The RT-ECR was built by Timothy Antaya. He received his PhD in Accelerator Physics from MSU in 1984, took a position at MSU/NSCL, and was later promoted to Senior Physicist; he left the laboratory in 1994. At NSCL, his main interest was in ion sources, and he built the RT-ECR, the first ECR source at NSCL. It made possible the non-coupling approach used in Phase II for the K1200. Later, he built the superconducting SC-ECR.

139. F. Marti and A. Gavalya, "Axial injection in the K500 Superconducting Cyclotron," Proceedings of the Eleventh International Conference on Cyclotrons (Tokyo, October 1986).

140. Quoted from "MSU/NSCL Project History," by H. G. Blosser, dated April 1982.

141. A detailed account of the problem and its solution is found in Section IV.4 of MSUCL-552, a proposal to

the NSF dated March 1986.

142. Floating Point Systems FPS-164.

143. Proposal from NSCL to Harper Grace Hospitals, requesting funding for a joint project to construct a superconducting cyclotron system to provide a neutron therapy modality for clinical radiation oncology applications, MSUCL-440.

144. E-mail from Jay Burmeister, Chief of Physics, Karamanos Cancer Center Detroit, dated May 8, 2013.

145. Proposal to NSF from NSCL for a Manufacturing Prototype Superconducting Cyclotron for Advanced Cancer Therapy, MSUCL-874.

146. Personal communication from Emanuel Blosser, May 8, 2013.

147. Letter from Sam Austin and Henry Blosser to William Rodney, NSF, dated December 4, 1985. Copy to site visit Committee for Proposal to NSF for 1986 Operating Funds.

148. Recollections and notes of the author, Sam M. Austin.

149. The details are laid out in a letter from John Cantlon, VP for Research and Graduate Studies at MSU, to Harvey Willard at the NSF, dated May 1, 1986. It involved additional contributions to the project of $1.66M each from NSF and MSU, and a waiver of the overhead on increases in the operating grant over the present amount of $5.45M.

150. Section I of MSUCL-552, a proposal to the NSF dated March 1986, for a detailed discussion of this issue.

151. Section II.B of MSUCL-552, op. cit.

152. Sherrill presented these general ideas at a workshop at LBL in 1984 (see LBL18187, UC/34A/CONF 8404154). Craig Snow and Al Zeller were also deeply involved in the A1200 design.

153. E. Kashy, S. M. Austin, M. R. Maier, J. Yurkon, J. Winfield, D. J. Mercer, D. Mikolas.

154. The quality of the RCA final amplifier tubes provided by the manufacturer had declined, and the tubes had become unreliable. The amplifiers were modified to accept TH555 tubes from Thompson, which gave much more reliable performance. It had also become necessary to replace the rectification and control aspects of the radio frequency power supply resulting in the present "Phoenix system."

155. The K500 and K1200 could be regarded as complementary, the K500 being adequate for relatively low-energy beams, and the K1200 unique for higher-energy beams. Running both would allow more physics to be done. On the negative side was the additional cost of running two machines in manpower, consumables, and problems of coordination. Nor could operation of the K500 be expected to be reliable. Finally, the NSF was not sympathetic to requests for the additional funding necessary to operate both cyclotrons for physics experiments, which settled the issue.

156. Acceptance is the range of angles for reaction products which can be analyzed by the device; for the S800, it is seven degrees vertically by ten degrees horizontally. This acceptance is sufficiently large so that the important angles for reaction products are accepted. Thus the spectrograph has almost never (only once) been rotated away from zero degrees.

157. M. Berz et al., "Reconstructive correction of aberrations in nuclear particle spectrographs," *Phys. Rev. C* 47 (1993): 537.

158. MSUCL 685, Proceedings of the International Conference on Heavy Ion Research with Magnetic Spectrographs (January 1989), N. Anantaraman and B. M. Sherrill, eds.

159. Proposal for the construction of the S800 Spectrograph, MSUCL-694, July 20, 1989.

160. The initial optical design of the S800 and its analysis beam were due to Lee Harwood, Jerry Nolen, and Al Zeller. Brad Sherrill led the push to complete the S800 after Nolen left the lab. Other contributors were Steve Bricker, Jac Caggiano, Jon Dekamp, Len Morris, Dave Sanderson, John Vincent, and John Yurkon. Daniel Bazin developed the methods for particle trajectory reconstruction.

161. Report of the Subcommittee (chaired by John Schiffer) on the Implementation of the 1989 Long Range Plan for Nuclear Science, submitted to NSAC on April 6, 1992, and forwarded to NSF/DOE on April 15, 1992, with additional NSAC comments.

162. Report of the NSAC Subcommittee (chaired by Robert Redwine) on the NSF Sponsored National User Facilities for Nuclear Physics, February 1993.

163. The compact style of the K500 or K1200 could not be extended to these higher energies, and the separated sector conformation carried with it both technical problems, probably solvable, and high costs. A similar device has now been built in Japan at RIKEN.

164. C. K. Gelbke, private communication, October 2013.

165. Felix Marti received his PhD in Physics from MSU in 1977, spent two years on the faculty of the University of the Republic, Uruguay, and returned to MSU in 1979; he was promoted to Senior Physicist in 1990. He has made major contributions to development of accelerators at MSU and was responsible for the coupling approach used in the CCF. He was Head of Accelerator R&D Department (1993–2009) and is now Group Leader of the FRIB Charge Stripper and Transport Area. For his CCF ideas, see: F. Marti et al., "High Intensity Operation of a Superconducting Cyclotron," Proceedings of the Fourteenth International Conference on Cyclotrons and Their Applications (Cape Town, South Africa), 45.

166. IUCF Annual Report 2003.

167. D. J. Morrissey et al., "Commissioning the A1900 projectile fragment separator," *Nucl. Instr. Meth. B* 204 (2003): 90.

168. Richard York received his PhD from the University of Iowa in 1976, and in 1993 came to MSU/NSCL as Professor and Associate Director for Accelerator Physics (1993–2010). Following his PhD, he was involved in the successful CEBAF proposal and in the resulting project. Later he managed the low- and high-energy booster projects at the Superconducting Super Collider. He was responsible for construction of the CCF and introduced project planning to the NSCL. He later led the initial efforts to develop technology for superconducting linear accelerators.

169. P. D. Zecher et al., "A large-area, position-sensitive neutron detector with neutron/gamma-ray discrimination capabilities," *Nucl. Instr. Meth. A* 401 (1997): 329.

170. B. Luther et al., "MoNA—The Modular Neutron Array" *Nucl. Instr. Meth. A* 505 (2003): 33.

171. R. Ringle et al., "The LEBIT 9.4T Penning trap mass spectrometer," *Nucl. Instr. Meth. A* 604 (2009): 536.

172. These reasons include: the separation process is quick, so one can study short lifetimes; one can use much thicker targets; one can identify and study several nuclei at one time; and one can use the focusing in the forward directions natural to reactions with fast beams to collect a larger fraction of reaction products. The product of these factors is often large.

173. The IsoSpin Laboratory (ISL), "Research Opportunities with Radioactive Beams," LALP91-51, North American Steering Committee for ISL.

174. 1991: The IsoSpin Laboratory, ISL Steering Committee (report).
    1995: TUNL Town Meeting, January 19–21, 1995. Report endorsing ISOL.
    1995: Overview of Research Opportunities with Radioactive Nuclear Beams, An Update (ISL).
    1995: Argonne Yellow Book: ANL-ATLAS Exotic Beam Facility.
    1997: Columbus White Paper: "Scientific Opportunities with an Advanced ISOL Facility."
    1999: "Nuclear Physics: The Core of Matter, the Fuel of Stars," National Academy of Sciences Report.
    2000: White Paper: "Scientific Opportunities with Fast Fragmentation Beams from RIA."
    2000: Durham workshop White Paper on RIA Physics.
    2003: White Paper: "The Intellectual Challenges of RIA."

175. Brad Sherrill and Konrad Gelbke (ex-officio) were members of the Task Force. Richard York and Sherrill were members of the Driver Working Group; Felix Marti and John Vincent were consultants.

176. Letter from C. K. Gelbke (NSAC Chair) to Hermann Grunder, dated October 21, 1998. The report referred to was the product of a workshop held in Columbus, Ohio, July 30 to August 1, 1997.

177. Private communication from C. K. Gelbke, October 2013.

178. The option of reaccelerated beams from a PF facility had been considered as early as 1989 as part of discussions during preparation of the 1989 NSAC Long Range Plan. In the meantime, Guy Savard (ANL) had developed a concept for stopping the fast beams. (B. Sherrill, private communication, October 2013).

179. S. Schwarz et al., "The NSCL cyclotron gas stopper—Under construction," *Nucl. Instr. Meth. B* 317 (2013): 463.

180. D. Leitner, "State-of-the-art post-accelerators for radioactive beams," *Nucl. Instr. Meth. B* 317 (2013): 235. This is an up-to-date summary.

181. J. Cederkall et al., "REX-ISOLDE—Experiences from the first year of operation," *Nuclear Physics A* 746 (2004): 17c.

182. John Vincent et al., "On active disturbance rejection based control design for superconducting RF cavities," *Nucl. Instr. Meth. Phys. Res. A* 643 (2011): 11.

183. H. Schatz, private communication, December 2014.

184. Presentation to Ray Orbach, Peter Rosen and Dennis Kovar of DOE, by Konrad Gelbke, Richard York, and Howard Gobstein of MSU.

185. T. L. Grimm et al., Proceedings of the Eleventh Workshop on RF Superconductivity (Lübeck/Travemünder, Germany, 2003), 32. Also, R. C. York et al., Proceedings of LINAC (Knoxville, TN, 2006), 103.

186. Letter from Konrad Gelbke to Patricia Dehmer of DOE, who forwarded it to Dr. Jehanne Simon-Gillo, Acting Associate Director of the Office of Science for Nuclear Physics.

187. Compiled from data provided by C. K. Gelbke.

188. GRETINA, and similar detectors, could localize the energy deposited by gamma rays interacting and scattering in the detector. This made it possible to determine the direction of the incident gamma ray with high precision, a crucial property for working with gamma-emitting, high-energy ions such as those from the superconducting cyclotrons and eventually from FRIB.

189. A Proposed Heavy-Ion Cyclotron," Report of the Committee on Nuclear Research" (Ballam Committee), June 1957.

190. See M. Thoennessen, "Current status and future potential of nuclide discoveries," *Rep. Prog. Phys.* 76 (2013): 056301.

191. For references, see the sections on cross-disciplinary research in: NSCL Operating Proposals MSUCL-972 (March 1995), NSF-0110253 (September 2001), NSF-0606007 (November 2005), and NSF-1102511 (October 2010).

192. Raed A. Alduhaileb et al., *Responses of Carbon Onions to High Energy Heavy Ion Irradiation*, MSU Elect. Eng., NSCL collaboration.

193. V. M. Ayres et al., *Investigation of Heavy Ion Irradiation of Gallium Nitride Nanowires and Nanocircuits*, MSU Elect. Eng., NSCL collaboration.

194. Michael K Bowman, et al., *Track Structure in DNA Irradiated with Heavy Ions.*

195. Loro L. Kujjo et al., *RAD51 Plays a Crucial Role in Halting Cell Death Program Induced by Ionizing Radiation in Bovine Oocytes,* PPNL, MSU Physiol., NSCL collaboration.

196. W. K. Kwok et al., *Anisotropically Splayed and Columnar Defects in Untwinned $YBa_2Cu_3O_{7-\delta}$,* ANL, NSCL collaboration.

197. W. K. Kwok et al., *Vortex Pinning of Anisotropically Splayed Defects in $YBa_2Cu_3O_{7-\delta}$,* ANL, NSCL collaboration.

198. Brian Geist et al,. *Radiation Stability of Visible and Near Infrared Optical and Magneto-Optical Properties of Terbium Gallium Garnet Crystals,* MicroXact, NSCL collaboration.

199. A. Pen et al., "Design and construction of a water target system for harvesting radioisotopes at the National Superconducting Cyclotron Laboratory," *Nucl. Instr. Meth.* A747 (2014): 62.

200. See: www.energy.gov/mission.

201. R. Zegers, Presentation, June 2013 and June 2014; NSCL Graduate Studies Brochure for 2013.

202. R. Zegers, op. cit.

203. Figures prepared by Michael Thoennessen and Remco Zegers, op. cit.

204. P. Apard et al., "Production of Multiply Charged Xenon Ions," *Physics Letters* 44A (1973): 432.

205. G. Machicoane, August 8, 2014, private communication.

206. The S320 is described in NSCL Annual Report for 1979–80: J Nolen et al., "Design and Construction of a K = 320 Spectrograph for Phase I," 72–73. See also: B. M. Sherrill, PhD thesis, Michigan State University, 1984, which lists other contributors W. Benenson, S. Bricker, W. Lynch, J. van der Plicht, D. Swan, J. Winfield, and J. Yurkon.

207. L. H. Harwood and J. A. Nolen, "A Reaction-Product Mass Separator for Energetic Particles at MSU/NSCL," *Nucl. Instr. Meth.* 186 (1981): 435. See also NSCL Annual Reports from this era.

208. The 4π Array's design and construction was led by Gary Westfall and had an innovative electronic setup designed by Michael Maier. Its detectors consisted of a thin and thick plastic scintillator (phoswich) for light ions, followed by a Bragg Curve detector for intermediate masses and a thin avalanche detector for heavy particles.

209. The Miniball's construction was led by R. T. de Sousa. Its 188 CsI + plastic phoswich detectors gave element identification up to Z = 18.

210. Built by Fred Becchetti of the University of Michigan with funds obtained from the DOE; it was commissioned in March 1992.

211. W. U. Schroeder, "Superball: A 4π Neutron Detector for Calorimetric Studies of Intermediate Heavy-Ion Reactions," 1993.DOE/ER/79048-1, DOE Performance and Final Report, March 1995. The Superball was completed in 1993.

212. See MISTIC on Wikipedia.org.

213. J. O. Kopf and P. J. Plauger, "JANUS: A Flexible Approach to Real-Time Timesharing," Proceeding AFIPS, 1968, Fall Joint Computer Conference, part II: 1033–42.

214. Done with the program TOOTSIE, D. M. Bayer, 1969 (unpublished).

215. Gary Westfall, private communication.

216. Quadrupole magnets are used to focus or reduce the size of a particle beam, much as glass lenses focus light. The process is, however, somewhat more complex. A quadrupole squeezes a particle beam in only one dimension and expands it in the other. Thus a round beam becomes an oval beam. Then, after passing through a second quadrupole, oriented to squeeze the now long axis of the oval, the beam is again round, but with a smaller radius. To focus, quadrupole magnets are often used in doublets or triplets.

217. K for Kelvin, a unit of temperature which is zero at absolute zero (–273.15°C). So room temperature is about 293 K.

218. In this discussion, I have, for simplicity of presentation, assumed that all the capacity of the liquefiers is used to produce liquid helium. In practice, a substantial fraction of liquefier capacity is used to cool the warm gas returned from the batch-filled magnets, so this cost is a lower limit.

219. A description of this system can be found on the MSU Human Resources website, currently www.hr.msu.edu/documents/facacadhandbooks/nsclhandbook/.

220. Letter from C. L. Winder to Henry Blosser, dated November 18, 1980.

221. Letter from H. G. Blosser to Provost C. L. Winder, dated October 1, 1984, describing their discussion of August 16.

222. A description of this system can be found on the MSU Human Resources website, currently www.hr.msu.edu/documents/facacadhandbooks/NSCLFacPos.htm.

223. C. K. Gelbke, private communication, April 13, 2015.

224. Excerpted from the description of the Tom W. Bonner Prize (2004).

225. Excerpted from the description of the Tom W. Bonner Prize (2012).

226. B. Sherrill, J. Bailey, E. Kashy, and C. Leakeas, *Nucl. Instr. Meth.* B40/41 (1989): 1004.

NOTES ON SOURCES

FIGURES

The source of most figures in this book is Michigan State University, and figures not specifically listed below are credited to: © Michigan State University.

Figure 2. Kurson, License: https://creativecommons.org/licenses/by-sa/3.0/legalcode.

Figures 3, 5. © 2010 The Regents of the University of California, Lawrence Berkeley National Laboratory.

Figure 6. *Lawrence and His Laboratory: A History of the Lawrence Berkeley Laboratory, Volume I,* by J. L. Heilbron and Robert W. Seidel, © 1991 by the Regents of the University of California. Published by the University of California Press. Reprinted with permission of the publisher.

Figure 11. MSU Academic Computing and Network Services.

Figures 13, 29, 35, 49, 65. © 1966 IEEE. Reprinted, with permission, from David L Judd, *IEEE Transactions on Nuclear Science,* 13 (4) (1966).

Figures 7, 8, 9, 15, 23a, 23b, 23c, 23d, 23f, 37, 41, 42, 50. Michigan State University Archives and Historical Collections.

Figure 25c. *FRIB: Opening New Frontiers in Nuclear Science,* August 2012, prepared by FRIB User Organization (FRIBUO).

Figure 26. Martin Savage, Institute for Nuclear Theory, University of Washington, Seattle, WA.

Figure 27. Reprinted with permission from American Physical Society, Copyright © 1968, American Physical Society.

Figure 30. Jerry A. Nolen.

Figure 51. Reprinted with permission from American Physical Society, Copyright © 1984, American Physical Society.

Figure 56. TRIUMF, Tri-University Meson Facility, Vancouver, BC, Canada.

Figure 72. C. K. Gelbke.

Figure 82b. *"FRIB: Opening New Frontiers in Nuclear Science,"* August 2012, prepared by FRIB User Organization (FRIBUO).

Figure 83. David Waymire.

TEXT MATERIAL

Some of the information in the first part of this book follows closely the material in "The Michigan State University Cyclotron Laboratory: Its Early Years," Sam M. Austin, in process of publication in *Physics in Perspective.*

THANKS AND APPRECIATION

This book's origin was an attempt, for my personal satisfaction, to recall the important events of fifty years spent at MSU and the MSU Cyclotron Laboratory. In a sense it was to replace the informal journal I never kept while events were still, mostly, accessible to memory. Some aspects of this work have been previously prepared for *Physics in Perspective*. Then, with the decision that the laboratory would celebrate the fiftieth anniversary of first beam from the K50 cyclotron, a more formal presentation seemed desirable and has led to this book.

Of course, many people helped turn it from a concept to a reality. I am greatly indebted to MSU President Lou Anna K. Simon for her perceptive foreword, expressing her belief in the role of a university in pushing forward the frontiers of both science and society. The expertise, guidance, and support of Julie Loehr and Gabriel Dotto at MSU Press was invaluable, as was the support of Senior Associate Vice President Paul M. Hunt. Sandra Conn played a crucial role in reshaping and refocusing the presentation and pointing out where a less-technical approach was needed. She also facilitated arrangements with the MSU Press to organize the necessary technical assistance for editing and designing the book and for funding from MSU. When Sandra retired in June 2015, Jane Miller took over these tasks.

Konrad Gelbke read a relatively early version of the manuscript and parts of the almost final version in some detail and helped clarify and sharpen the presentation. Thomas Glasmacher and Brad Sherrill read the final document in its entirety and provided valuable additional input and corrections. Finally, many laboratory people, and especially Erin O'Donnell and Don Lawton, provided photos and graphics. Felix Marti advised me on many technical issues. And Karen King helped with discussions of the many illustrations and of the nature of appropriate credit. Mary Austin proofread the entire book with great care.

Editing and designing the book fell to Alicia Vonderharr and Charlie Sharp; their willingness to work within a greatly accelerated production schedule, and their care and skill in the book's production are greatly appreciated.

All of these individuals have my thanks for their contributions. I greatly appreciate, also, the patience of those, especially my wife Mary Austin, who allowed me to neglect other tasks to finish this project, and who did in their entirety many tasks I should have shared.